Ranch Roping

Also featuring Buck Brannaman:

Believe: A Horseman's Journey
and
*The Faraway Horses: The Adventures and Wisdom
of One of America's Most Renowned Horsemen*

Both by Buck Brannaman with William C. Reynolds
and published by The Lyons Press

Ranch Roping

The Complete Guide to a Classic Cowboy Skill

Written by Buck Brannaman and A. J. Mangum

Photographed by A. J. Mangum

THE LYONS PRESS
Guilford, Connecticut
An imprint of Rowman & Littlefield

Copyright © 2009 by Buck Brannaman and A. J. Mangum

ALL RIGHTS RESERVED. No part of this book may be reproduced or transmitted in any form by any means, electronic or mechanical, including photocopying and recording, or by any information storage and retrieval system, except as may be expressly permitted in writing from the publisher.

The Lyons Press is an imprint of Rowman & Littlefield.

Distributed by NATIONAL BOOK NETWORK

Library of Congress Cataloging-in-Publication Data

Brannaman, Buck.
 Ranch roping : the complete guide to a classic cowboy skill / Buck Brannaman and A.J. Mangum ; photographed by A.J. Mangum
   p. cm.
 ISBN 978-1-59921-447-4
 1. Ranch roping. I. Mangum, A. J. II. Title.
 SF88.B72 2009
 636.2'083—dc22

2008028261

Printed in the United States of America

Contents

Introduction
Vaqueros, Buckaroos, and Buck Brannaman — vi

Chapter One: The Art of Roping
Ropes — 1
Getting Started — 4
Forward Swings — 9
The Houlihan — 22
The Backhand — 25

**Buck Brannaman on
Ranch Roping's Heritage** — 39

Chapter Two: The Mechanics of Dallying
Dallying — 43
Logging — 52
Tracking — 63

**Buck Brannaman on
Ranch Roping's Horsemanship Challenges** — 68

Chapter Three: The Cowboy and the Herd
The Rodear — 73
Working Scenarios — 75
Three-Man Doctoring — 90
Two-Man Doctoring — 98

**Buck Brannaman on
Strategies for Beginning Ropers** — 107

Chapter Four: Tools for Training
Headgear — 111
Hobbling — 121

Credits — 124

Introduction

Vaqueros, Buckaroos, & Buck Brannaman

A guy's going to get the job done, but he shouldn't have to apologize for enjoying it while he gets the job done.

— BUCK BRANNAMAN

As with many words and expressions in the cowboy vernacular, the term *roping* can take on more than one connotation. For the rodeo or jackpot competitor, roping is a timed event that takes place inside the confines of an arena.

In team roping, calf roping, and breakaway roping, speed is everything. A competitor's horse must break from the box at just the right instant, and the roper must make his catch in the shortest time possible.

For ropers in such competitive events, a brand of highly technical horsemanship is at work, one built upon achieving machinelike efficiency: getting a horse calm and collected inside the box, perfectly timing the cues that send the horse surging forward without breaking the barrier and incurring a penalty, positioning the horse so that a catch can be made, and through it all, handling a rope safely and with precision, from swing to throw to catch to dally.

The term *ranch roping,* however, refers not to a rodeo or horse-show arena event but to the practice of roping cattle on the open range, or in a ranch corral, in order to restrain them for branding or doctoring. On a ranch, roping is not a sport meant to entertain a crowd of spectators or feed the competitive desires of its participants. It is a necessary skill, one with a practical purpose.

In working scenarios, ropers are judged not by the speed of their performances but by the quality of their work—the accuracy of their catches, the efficiency of their movements, their adherence to sound horsemanship practices, and their ability to work quietly and limit the stress they place on cattle.

For the working cowboy, a good performance is not rewarded with a buckle or trophy or prize money, beyond the wages he might be owed by the rancher employing him. In Zen-like fashion, a cowboy's reward lies in the job itself: the successful completion of a necessary task, such as branding a calf or vaccinating a sick cow; the humane treatment of cattle, with respect for the fact that

Buck Brannaman's working methods descend from those of the California vaquero and Great Basin buckaroo.

they are the source of the rancher's livelihood and, at least indirectly, the source of the cowboy's; the satisfaction of working in partnership with fellow cowboys, of being in the right place at the right time, and knowing that he can count on that same support; and an adherence to a strict code of horsemanship, one based on centuries of tradition and centered on the ethical treatment of the cowboy's closest working partner.

It all adds up to the ultimate reward: earning the right to be a trusted member of a working cowboy crew and deserving of the privilege of returning the next day. To a working cowboy, that is worth more than any buckle or trophy.

The California vaquero style of ranch roping dates from the eighteenth century, when Franciscan missionaries first arrived on the Pacific Coast, bringing with them small herds of cattle, the seed stock for the vast *ranchos* that would thrive under Spanish rule. The vaqueros were the ranch cowboys of Spanish California, and, as horsemen, occupied a higher social caste than that of the common laborer.

Introduction

Known for his impeccable horsemanship, high standards for horseflesh, and appreciation for finely crafted, even ornate, working gear—intricately braided rawhide and elaborate, silver-adorned bits and spurs—the vaquero took equal pride in his stock-handling talents, emphasizing the skill to effectively handle cattle from horseback and the ability to do so with style and elegance, whether herding cattle or roping with a rawhide *reata,* the vaquero's rope of choice.

In 1850 California joined the Union, and the vaquero culture spread inland to what would become Nevada, southern Idaho, and southeastern Oregon. In this rougher, less-forgiving country, the vaquero evolved. The term itself—*vaquero*—became Anglicized into *buckaroo.* And, whereas the California vaquero had been a high-profile character, the Great Basin buckaroo adopted a quieter, more reclusive nature, with some populations barely registering on censuses of the time.

The buckaroo, however, inherited the vaquero's appreciation for refined horsemanship and stock-handling skills, working gear that was functional and stylish, and a preference for using a rawhide reata to rope livestock.

Detailed knowledge of buckaroo working techniques, including the ways they trained their horses and handled the reata, has not always been easy to come by. Members of the buckaroo culture tended—and to a certain degree, still tend—to value solitude over socialization.

Trade secrets related to starting colts, handling stock, and working a rope set a buckaroo apart from his peers, and went a long way in defining his very identity, both as an individual and as a proud member of the brotherhood of Great Basin stockmen. Such information was typically passed down only from father to son, or on rare occasions, shared with select individuals a buckaroo deemed worthy of teaching.

Little was written about how a buckaroo worked, and well into the twentieth century, there was minimal effort at publicizing or sharing with the rest of the world the methods of this particular subculture of horsemen.

That began to change in the middle of the twentieth century, when a revolution in horsemanship began. Horse owners, many of them among the growing population of recreational riders without the benefit of ranch experience, began searching for better, more effective ways of working with their horses. They sought solutions to training challenges that seemed impossible to overcome, and hoped to discover horse-handling techniques that were gentler and more humane than the mainstream methods of breaking and training they had come to view as harsh, violent, and cruel.

Word of mouth led horse owners from across the United States to two brothers, Bill and Tom Dorrance, veteran Great Basin cowboys whose near-mythic reputations as horse handlers had been forged on cattle outfits from northeastern Oregon and northern California to the remotest stretches of the Nevada outback. Each brother seemed to possess an innate extrasensory awareness of a horse's emotions, reactions, and interpretations of human behavior.

The two men emphasized the need to understand the horse's mind-set, and taught cowboys to work *with* their horses, rather than simply imposing their will upon the animals through brute strength.

Ray Hunt was the first horseman to bring the Dorrance philosophies to the rest of the world. Forty years after he began conducting clinics, Hunt's methods are still considered the standard by which all work with young horses is measured.

Born in 1962, Buck Brannaman grew up in northern Idaho and central Montana. During his childhood, he performed as a trick roper at rodeos, on television programs and commercials,

Buck Brannaman spent his childhood performing as a trick roper and working cattle on the Montana ranch where he was raised. Roping has been a part of his daily routine for nearly all his life. A protégé of master horsemen Ray Hunt, Bill Dorrance, and Tom Dorrance, Brannaman has spent the last two decades conducting clinics on horsemanship, colt starting, and ranch roping. One of today's most respected horsemen, Brannaman is recognized as one of the world's best hands when it comes to handling a rope in working situations.

and during his teen years, as part of a U.S. State Department–sponsored goodwill tour that took him from the United Kingdom to Japan.

By the time Buck began cowboying on Montana ranches as a young adult, he could make a rope behave as if it were anything but a twisted and braided length of inanimate fibers, as if it were instead simply an extension of his hand, an extremity over which he had complete control.

Today, Buck is one of the world's most respected horsemen. A protégé of Ray Hunt and of both Dorrance brothers, Buck has spent much of the past two decades on the road, conducting horsemanship, colt-starting, and ranch-roping clinics throughout North America, Europe, Australia, and New Zealand, teaching smarter, safer, and more effective ways of working with horses and achieving stronger partnerships with them.

Buck and I first teamed up on a ranch-roping project in 2000. The result was *Ranch Roping with Buck Brannaman,* a forty-page booklet covering some rope-handling fundamentals and the mechanics of about twenty shots used in working situations by ranch cowboys.

When we began that project, roping was not new to me. I had grown up on a horse and cattle ranch in central Oregon and had competed in Western events for many years. I had been around roping, and ropers, all my life.

Working aboard a bridle horse at his ranch outside Sheridan, Wyoming, Buck sends a long-distance loop into the air. After a lifetime of rope handling, Buck has the ability to make a rope perform according to his will, a talent he puts to work on his own ranch and on other working outfits throughout the West. At his clinics—around forty each year—he teaches his students to handle a rope safely and efficiently when working cattle.

I understood the idea that roping was about more than the sum of its parts—the cowboy, rope, horse, and cow—and that for ropers, the mechanics of their craft fueled what can only be described as an addiction, and with continual improvement, a lifelong obsession.

Roping, I had observed, had a way of changing the lives of newcomers to the discipline. People once firmly committed to the idea of a good night's sleep would begin staying up until the wee hours of the morning, throwing hundreds of practice loops at plastic steer heads attached to sawhorses or bales of hay. They would start carrying ropes with them everywhere they went so their new tools would be nearby and easily accessible on the off chance some opportunities arose to throw a few loops.

And, once they progressed to roping on horseback, only the threat of personal disasters of epic proportion could keep them from loading their horses in their trailers and heading to the weekly jackpot, or to a local ranch that needed some extra help at its branding.

Working cowboys, I knew, intently studied their fellow ropers' techniques—hoping to pick up on an idiosyncrasy or variation that might be worth adopting—and shared in a brand of camaraderie that did not seem to exist inside the competitive arena. It had been my experience that when a working cowboy sent a loop into the air, his compatriots all hoped his rope found its mark, and when a catch was made, everyone present shared in the momentary thrill, even if only vicariously. If jealousy of another's skill

Introduction

existed, it did so only in the context of motivating the lesser roper to improve.

Long story short, I thought I had more than a solid read on the roping culture. However, until I traveled to Buck's Wyoming ranch to shoot photographs for the booklet and, for the first time, watched him swing a rawhide reata, I had never thought of roping as an art form.

With some luck, any task born of necessity can be executed amateurishly, with little or no finesse, and still produce passable results. But when a person takes seriously the execution of that skill and chooses to devote a significant piece of his or her life to improving and learning smarter techniques, efficiency increases and elegance begins to emerge. A workaday task becomes appreciated as much for its processes as for the result it produces.

This, I learned, is Buck Brannaman's mindset in his approach to roping, and to much about life in general.

During that first visit to Buck's ranch, as I watched him rope he displayed a level of skill attainable only by someone with an unappeasable obsession with roping and a lifetime of experience putting his skills to work with stock. When he built a loop and began to swing, Buck seemed to transform, as an actor might as he walks onstage and shuts out all distractions, immersing himself completely in his character.

Once Buck locked on a target and released his loop, all eyes turned toward him. It was plain to everyone present that Buck's every sense was tuned in to just three things: his horse, his rope, and the cow. There was nothing else.

The moment Buck's loop left his hand, had mind reading been among my few talents, I would have expected to share the sensory experience of his locked posture and suspended breath, and the feeling of the reata's texture as it left his hand. I would have anticipated a set of mental images depicting the catch Buck expected to make, the catch that was about to occur—the rope sailing through the air, often across what had seemed an impossible distance for it to be thrown, the coils playing out, the loop opening and taking shape as it gracefully settled over the cow's head.

When Buck would make his catch, pull his slack, dally, and "face up" to the cow, the slow-motion sequence of the preceding moments would accelerate to real time, and once again he would become part of the world around him, seemingly no longer immersed in an insulating zone, but instead hyperaware of his surroundings, ready to react to every movement of the cow and anything that could jeopardize the task at hand or pose a risk to himself, his horse, or those around him.

The specific scenario never seemed to matter. Buck could be heading or heeling, working with partners or working alone. The cow might be in front of Buck, behind him, embedded in a tightly clustered herd, or on the move in any given direction. Still, Buck's rope always seemed to find its target. It was the closest thing to telekinesis I had ever witnessed.

Like those cowboys whose greatest reward is the privilege of doing the job, I felt lucky to be where I was, watching a master at work, and trusted with the formidable task of documenting Buck's approach to a beautiful craft, a classic cowboy art.

I would be guilty of an appreciable degree of negligence if I did not address up front, however reluctantly, the limitations of the written word when it comes to the subject at hand.

Roping is not a skill to be learned from a set of instructions; it must be learned through doing. Therefore, this volume should not be regarded as an instruction manual. Rather, this is a collection of lessons, insight, and wisdom on the topic of ranch roping, a work meant to capture the imagination of the beginner, inspire the novice, and provide clarity for the veteran.

Buck has just one coil remaining in his left hand as a long throw approaches its mark. And, yes, he made the catch.

His dallies in place, Buck celebrates a long catch as Kevin Hall (in the background, on the right) moves into place to heel the captured heifer.

In the following pages, beginning ropers will learn the fundamentals of rope handling and ways to marry those skills with proper horsemanship, itself a vital component of ranch roping. Budding ropers with limited experience will acquire a clearer understanding of some of roping's finer points and perhaps find the encouragement to improve their skills by throwing thousands of loops on a roping dummy and thousands more on live cattle, under the watchful eye of a more seasoned hand.

And, those seasoned hands will find affirmation of many of the techniques they have used for decades, and perhaps learn a few new tricks as well.

The instructional sections of this book have been organized in what we hope will prove to be a logical sequence, so that each stage of learning draws upon the lessons of the previous stages. Given the technical nature of the material to come, some suggestions for a reading strategy might be in order.

Any instructional text calls for more than one reading. After an initial read-through of a particular section, you might benefit from an immediate second reading, this time with more

Introduction

active visualization of any horseback maneuvers described and, of course, any arm and hand movements that are outlined.

Before moving on to subsequent chapters, you might step outside and put to the test the lessons you have learned. Afterward, a third reading of the relevant text might be in order, as the text might after practical application take on a new meaning.

Ideally, you should have a partner equally interested in roping, someone with whom you can share your interpretations of this volume. You can critique one another's roping, offer reminders of proper technique, and in the spirit of camaraderie that is a part of the ranch-roping culture, keep each other motivated.

On the specific topic of dallying, one can't overemphasize the need for extreme caution. The cowboy population includes a good number of ropers with fewer than ten fingers. As anyone unlucky enough to be part of that particular demographic will tell you, improper dallying can lead to serious, permanent injury.

When you begin to practice dallying, do so under the supervision of a veteran roper who can serve as a spotter, someone who can keep track of your hand positioning, offer critiques, and warn you if you are putting your digits in danger.

One last note: For the sake of simplicity and consistency, this book's guidelines for rope-handling have been written with right-handed ropers in mind. Left-handed ropers should reverse the text's right-hand and left-hand cues.

The nature of roping is such that, the more you do it, the clearer its mechanical processes become. But like horsemanship, roping is a lifelong pursuit, one that continually reveals new challenges and new lessons, and offers opportunities for improvisation and the development of one's own techniques. There is no end point in learning this skill. There is no singular goal to be reached. Rather, this art and craft revolves around a process and a philosophy of ongoing improvement.

That being said, let's begin.

—A. J. Mangum

Chapter One: The Art of Roping

The reata has a life of its own. Some days, it can confound you. Other days, it can make you look handy.

—BUCK BRANNAMAN

Ropes	1
Getting Started	4
Forward Swings	9
The Houlihan	22
The Backhand	25

Ropes

Among the tools of the cowboy's trade, the rope possesses a unique iconic status. A rider need not be a working cowboy to carry a rope on his saddle, but a working cowboy's rig is incomplete without one. And while it might say little about its user's skill level, a well-worn rope conveys much about the kind of work a cowboy does, affirming that his job is not limited to working with horses or riding fence lines and that much of his work revolves around cattle.

A working cowboy's rope is typically constructed of one of three materials: nylon; polypropylene, or poly; or in the case of a reata, untanned cowhide—also known as rawhide.

The hand-braided rawhide reata is intrinsically linked to traditional California-style ranch roping. Vaqueros in early California, lacking the maguey cactus fibers used by their compatriots in Mexico to fashion ropes, instead braided thin strands of rawhide to create a tool that is still a part of the contemporary buckaroo culture.

"The reata has its advantages and disadvantages," Buck says. "A brand-new reata feels awkward and stiff, but once it is broken in, it can

Clockwise from lower left: A traditional, handmade rawhide reata, the preferred rope for those adhering to the vaquero traditions of Spanish California; a poly rope with a rawhide honda made by master horseman and rawhide braider Bill Dorrance; a nylon rope with a metal honda; a poly "knot rope" with a metal honda; and a poly rope with a breakaway honda. In practice sessions, a breakaway honda, one that is notched so the loop breaks free of the cow after the roper makes his or her catch, allows for quicker retrieval of the rope, and a shorter set-up time before the next shot. Because of the craftsmanship that goes into making a hand-braided reata, and the cost attached to such work, many cowboys and buckaroos—even die-hard traditionalists—trade their reatas for poly or nylon ropes in sloppy working conditions.

[1]

One method of breaking in a new reata involves pulling it, from horseback, through a series of holes drilled into a post. Here, the post has been secured to the rail fence of a pasture, and the reata has been laced through four holes in the post. A rider can then dally on to his saddle horn and ride away at a slow walk to pull the reata through the series of holes. The rider would then lay the reata down, return to the post, and pull the reata through in the other direction. A stiff, new reata can be made smoother and more pliable after being pulled through the post several times. Before attempting this, make sure you "unroll" the reata so no coils go through the post.

The parts of a loop include the base, which is the top one-third of the loop; the honda, the eye the rope passes through to form a loop (seen here at the middle of the loop, on its left side); the spoke, the section of rope that is between the honda and the roper's throwing hand but not part of the loop itself; the tip, the bottom third of the loop; and the coils, the remainder of the rope held in the roper's non-throwing hand. Here, a head catch has been made, and Buck prepares to make a heel throw.

feel like an extension of your hand. It has a lot of weight to it, and because of that weight, if you are going to rope in the wind, there is nothing better. You can power through the wind and still throw long shots."

Most reatas are four-plait, handcrafted by braiding four rawhide strands, although more intricately braided six- and eight-plait reatas are not uncommon.

"A rawhide reata is remarkably strong, especially when it has been braided tightly, but it is still easy to break one, especially if you do not know how to run rope [feed slack] over the saddle horn," Buck explains. "Because of that, the reata is really only for skilled hands."

As an alternative, most ranch cowboys opt for less-expensive poly or nylon ropes for working situations. Such ropes are easier to replace and lack the sentimental value of a handcrafted reata.

"When you are roping on a ranch, a lot of times you are doctoring calves in sloppy, wet, miserable weather," Buck says. "You do not want to put a good reata through that. Manufactured ropes—nylons and polys—are plenty functional."

Because of its weight, a poly rope shares some of the reata's throwing characteristics but can easily rope-burn a bare hand if a working situation goes wrong.

In contrast, a nylon rope is lighter than a reata or poly, and might not throw as far, but can be more forgiving in that it will not burn a hand as badly as a poly rope might. This makes a nylon rope ideal for a novice roper.

Getting Started

The best place for a novice to begin work on the fundamentals of rope handling is on the ground, with no horse or cattle on the scene to create complications at this early learning stage. Likewise, the ground can be the best setting for a more experienced roper to practice the basics.

BUILDING A LOOP: Begin with the coils in your left hand and the beginning of a loop—essentially just the front coil and the honda—in your right hand. Grip the spoke of the loop in a closed fist with your fingers under, and thumb over, the spoke. Extend your right arm to your side, and let the loop hang straight down.

Next, flip the loop backward, over your right wrist, in what will appear to you a clockwise motion. As you do this, release a single coil from your left-hand grip, and with your right hand, allow the honda to slide up the spoke. The extra rope from that single released coil will feed into the honda, increasing the loop's size.

Now, momentarily shift the loop to your left hand, gripping the loop and the coils. With your right hand, take a new grip on the spoke of the loop, again making a fist with your fingers under the spoke and your thumb over the spoke.

Repeat the right-hand flip, increasing the loop's size. As you get more proficient at building a loop, these individual steps will blend into one motion.

A note on loop size: When on horseback, if a roper holds out the right arm at ninety degrees, the tip of the loop should be around twelve to eighteen inches from the ground. When practicing on the ground, a slightly smaller loop might be more practical. ❧

1. Begin building a loop by holding the base (the top one-third of the loop) and the spoke (the section of rope that is between the honda and the roper's throwing hand, but not part of the loop itself) in your right hand, with your thumb over the spoke.

Getting Started

2–4. Next, flip the loop backward over your right wrist. As you do this, release a single coil from your left hand. Allow the honda to slide up the spoke as the extra length of rope (from the coils) feeds into the loop, increasing its size.

Ranch Roping

5.

6.

7.

5–7. Shift the loop to your left hand, gripping both the loop and the coils. Take a new grip on the loop with your right hand, and repeat the right-hand flip depicted in the previous photo sequence. Repeat this maneuver as needed to build your loop to an appropriate size for the shot you are preparing to make. Here, Buck works with a traditional, hand-braided rawhide reata.

Getting Started

THE FIGURE-EIGHT DILEMMA: A kinked loop, one with a figure eight in it, can confound novice ropers and experienced hands alike. Rather than recoiling and then rebuilding the loop from scratch, hold the loop off the ground and examine it. To remove the figure eight, the loop simply needs to be rotated in one direction or the other.

Take hold of the base, the top part of the loop itself, in your right hand. Holding the spoke stationary, turn the loop once to remove the figure eight. Obviously, if the figure eight has worsened, you have turned the loop in the wrong direction.

If your rope happens to have a swivel honda, in which the rope can turn independently, you can remove the figure eight simply by turning the spoke inside the honda.

1–3. To remove a figure eight, rotate the loop in one direction or the other as you hold the spoke stationary. If the figure eight worsens, you have turned the loop in the wrong direction.

[7]

Ranch Roping

SORTING COILS: As you work, it is important that you keep the coils of your rope organized and of even size. Even in the most tranquil working situations or practice sessions, coils can easily become crossed and out of position in your left hand, meaning they will not feed cleanly and can become tangled, creating the potential for a serious horseback wreck, or at minimum, making throwing or dallying impossible.

"I am very particular about how even my coils are before I build a loop, and that the coils are the same size and organized," Buck says. "Before I start to work a cow, I will sort through my coils to make sure they are straight, and that none of them are crossed. That can be dangerous when you have a mad cow on the end of your rope."

As a rule, never allow yourself to be rushed into a working situation when your rope is not ready to be put to use safely. It is better to miss an opportunity for a catch than to put yourself, your horse, your partners, or the cow at risk because of a tangled rope that can't be handled effectively. The best solution to this problem: Drop the whole works. Take the time to recoil your rope entirely. Keep your new coils a uniform size, and as you work fix any kinks or figure eights.

Joel Eliot sorts through his coils—ensuring they are organized, and of the same size and shape—before he resumes roping. This important safety measure ensures that the rope is ready to be put to use, and that the coils will feed cleanly. Should coils become crossed, it not only inhibits the roper's ability to handle a job but also creates a potentially dangerous situation in which the roper or his horse can become tangled in the rope. When working cattle in a fast-paced environment, this could lead to a serious injury.

Forward Swings

To anyone who has observed ropers at work or in competition, the forward swing is the most familiar; from the roper's perspective, the loop swings in a clockwise direction.

A forward swing takes on one of three variations: a sidearm swing, with the loop in motion on the roper's (and horse's) right side; an overhead swing, with the loop in motion, as the name implies, above the roper's head; or an overhand swing, with the tip of the loop traveling overhead but angled downward over the roper's left shoulder. With subtle changes in arm or elbow positioning, a roper can shift from one forward-swing variation to another, allowing him or her to adjust as a working situation evolves.

Again, the best place to work on the fundamentals of these basic swings is on the ground.

SIDEARM SWING: To begin a sidearm swing, hold the loop in your right hand, gripping the spoke and the top of the loop loosely in your fist. The spoke should be deeper into your hand than the loop itself and held so it lies between the first and second joints of your index finger.

"Curl your fingers around the loop," Buck says, "and that is your basic position for holding the rope."

With the coils in your left hand, hold the loop parallel to your right side, with the loop hanging down, and swing the rope so the tip of the loop rolls forward, up, then back, making from your viewpoint a clockwise circle.

"When you are swinging a rope from the ground, always take into account where your horse's head would be if you were on horseback,"

1. The basic position for holding a loop: the spoke should be deeper into your hand than the loop, and held so it lies between the first and second joints of the index finger.

2. The sidearm swing should travel at around forty-five degrees, angling inward as it travels up, and outward as it travels down.

3. The sidearm swing is particularly useful for making head or heel catches on a cow traveling left to right in front of the roper. Eyeing a long-distance catch, Buck has "dropped coils," adding them to his right-hand grip so that when he releases his loop, it will easily travel over a longer distance. Here, Buck is astride a bridle horse outfitted in a spade bit and pencil bosal, and is swinging a traditional rawhide reata. He guides the horse with rawhide romal reins.

Buck says. "And even when you practice on foot, place your feet square on the ground, just as they would be if you were in the saddle with your feet in the stirrups. Hold your coils straight and forward of your body; that way, when you are in the saddle, you will not accidentally turn your horse right or left."

Continue swinging the rope, using a good deal of arm movement and rotating your shoulder. The loop should be wide open, with the angle of the sidearm swing at about forty-five degrees, angling inward as the loop travels up, angling outward as it travels down. If a figure eight forms in your loop, it is likely that you are using too much wrist movement in your swing.

"At no point in the swing should you see the palm of your hand," Buck points out. "If the palm of your hand turns up, the loop figure-eights. It is a typical mistake."

Forward Swings

OVERHEAD SWING: To achieve an overhead swing, begin a sidearm swing as described in the previous section. As you swing, raise your elbow to elevate the loop and flatten its angle so the loop now travels overhead. Once you are on horseback, if your loop is the correct size and your right-hand position is correct, the loop's path should be such that the honda travels directly above the horse's head with each rotation of the loop.

1. For the overhead swing, a roper raises his elbow so that the loop travels above his head parallel to the ground. Even when practicing on the ground it is important to position your body as if you were on horseback. Here, Buck makes an overhead swing, and holds his coils just as he would if he were in the saddle.

2. In the saddle, Buck keeps his elbow raised, elevating the overhead loop, and keeping it flat as it travels.

OVERHAND SWING: Next, adjust the loop angle by reaching farther to the left with your right arm. The loop's tip should now angle downward over your left shoulder. When this swing is made from horseback, the swing at its lowest point will pass in front of the horse's left eye when a basic heel trap is thrown.

Practice shifting from the sidearm swing to a flat overhead swing, then to an angled swing with the tip of the loop over your left shoulder. Then work backward, shifting back to the flat overhead swing and then to the sidearm. This will help you learn to control your loop and instantly adjust the angle of your swing, allowing you to respond immediately to a change in a cow's position.

The overhand swing should travel so that the tip of the loop angles downward over the roper's left shoulder.

PRACTICING THE FORWARD SWINGS: Dedication to roping requires consistent, frequent practice, ideally on a daily basis. Before you begin roping live cattle, a roping dummy makes an ideal target

To begin practicing your forward swings, place a roping dummy—essentially a sawhorse fitted with a plastic steer head—in an open area, and step back ten to twelve feet. Build your loop and begin a forward swing, following the guidelines provided earlier, then flatten the swing's angle in front. When you are ready, make your throw at the roping dummy's head, releasing the loop as your right arm is fully extended toward the dummy. Follow through with your right arm, rolling your right wrist slightly counterclockwise as you finish the motion.

Work at getting the loop to settle over the dummy's head. If you miss the dummy, retrieve your rope, rebuild your loop, and try again. Develop a feel for the timing and energy required to properly release the rope and catch your target. Remember to keep the loop's tip down.

Once you rope the dummy, pull your slack back to your hip using your right hand, palm down. After you have pulled your slack, note the position of your right hand. Remember that, on horseback, your right hand will need to be in position to dally at this point.

Retrieve your rope, rebuild your loop, and continue practicing, repeating the above exercise with the overhead swing and then an overhand swing—again, this is a forward swing with the tip of the loop angled downward over your left shoulder.

Forward Swings

1–3. Buck makes a head shot on the roping dummy using a forward swing. He releases the loop with his arm fully extended and, once the loop settles over the dummy's head, pulls his slack, just as if he were in the saddle. Note that as Buck pulls his slack, he brings the coils forward with his left hand, again mimicking the motion that will be necessary on horseback.

4–6. Buck stands to the dummy's left side and uses an overhand swing to make a head catch. The overhand frequently comes into play when a rider needs to make a head shot on a cow traveling from the roper's right to the roper's left.

Forward Swings

7–9. The overhand head shot requires a longer spoke, giving you more power at the end of the shot. Swing with power, throw with ease, and let the rope work for you.

10–11. The sidearm head shot is thrown in a left-to-right situation, from a swing of about forty-five degrees. You have to be quick taking your slack, or you will "belly rope" the cow. This is always the case when your rope swings in the opposite direction of the cow's motion. For this reason, Buck normally prefers the houlihan in such scenarios.

Forward Swings

12–13. Here, the sidearm head shot is applied on this black heifer.

FEEDING ROPE: Once you are comfortable making forward swings and, thanks to your practice sessions, you begin to understand the hand and arm positioning that will be needed to make effective throws, you should begin practicing feeding rope into the loop.

When you are ready to make a swing, the loop should be large enough so that, when it is in motion, its weight helps you build momentum in your swing. Too small a loop will lack this size and weight, and your swing will lack both speed and strength.

If you realize your loop just is not large enough, feed additional rope into the loop by loosening your right-hand grip on the spoke and letting a few inches of rope feed into the loop as the rope travels in the forward part of the swing. To build your skills, it is a good idea to practice feeding rope into each type of swing you work on.

Learning to feed rope will also prepare you to keep your loop balanced in relation to the weight of the honda. Feeding rope requires trial-and-error experimentation to do well. Feeding too much rope into the loop can create a figure eight in the loop, rendering it useless. ❧

RETRIEVING YOUR ROPE: When a cowboy has missed a shot, or when his rope has been removed from a cow, perhaps by another cowboy who has finished doctoring the animal, that cowboy needs to retrieve his rope in an efficient manner to prep for the next catch he will need to make.

Buck spells out three methods for rope retrieval. He adds that, no matter which method is used, it is important for a mounted roper to keep his reins in front of him and in a neutral position while retrieving the rope to not unintentionally cue his horse to move left or right.

Method 1: Grip the extended rope with your right hand, palm up. Coil the rope by rolling your right hand counterclockwise, forming a new coil that you then add to the coils in your left hand. Try to keep each new coil the same size as the others; if your coils become disorganized, grip the tail end of the rope in your left hand, drop the rest of the rope, and recoil its entire length. "Hard and fast" ropers use this method, as they do not dally.

Method 2: Grip the extended rope in your right hand, this time palm down, in dallying position. Slide your right hand down the rope and roll the part of the rope that is between your hands into a coil by drawing the rope toward you, rolling it as if you were tightening a nut. Form a new coil and add it to those in your left hand.

Method 3: If you are in a hurry to build a new loop, you can retrieve your honda without recoiling. Grip the extended rope and, with a quick, snapping motion, simultaneously lift up and pull. The honda will lift off the ground and, if you have used the correct arm and wrist motion, come flying through the air toward you. With practice, you can catch the honda in your right hand and immediately begin building a new loop by holding the honda in your right hand, hooking the loop with your left thumb, then whipping the slack to the side, being careful not to hit your horse while loading the loop. ❧

Forward Swings

Method 1: 1–3. Buck grips the rope with his right hand, palm up, and rolls his hand counterclockwise to recoil. MARY BRANNAMAN PHOTOS

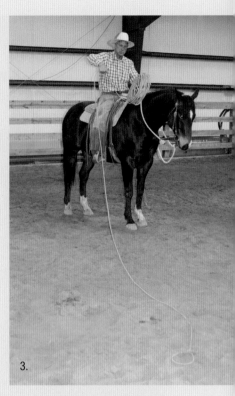

Method 2: 1–3. Buck demonstrates an alternate recoiling method, this time gripping the rope with his right palm down. He recoils by drawing the rope toward him and forming a new coil to add to those in his left hand. MARY BRANNAMAN PHOTOS

4. Kevin Hall uses Method 2 to recoil while working in a cattle pasture. His palm downward, he brings the played-out slack toward him, then rolls it upward and then away.

Forward Swings

Method 3: 1–4. Buck uses a quick, snapping motion to retrieve the honda. Having caught the honda, he rebuilds his loop. MARY BRANNAMAN PHOTOS

The Houlihan

Perhaps the most stylistically interesting of the basic swings, and one of the most useful when it comes to making long-distance catches, the houlihan is swung in the direction opposite that of the forward swing—traveling, from the roper's perspective, in a counterclockwise circle.

Swung at the roper's side, the houlihan travels at an angle similar to that of the sidearm swing—about forty-five degrees, with the loop angling inward as the tip travels up, and angling outward as the tip travels down.

Once again, practice the fundamentals of the new swing on the ground rather than on horseback. Remember to be mindful of the positioning of your feet and hands. Even though you are on the ground, work as if you were in the saddle, with your feet square as they would be in the stirrups and your arms positioned so that if you were on horseback, you would not unintentionally cue your horse to turn.

To swing the houlihan loop, begin by holding the loop and spoke in your right hand with your palm down so that you can see the back of your hand. Let the loop hang parallel to your right side.

Swing the loop so that the tip rolls back, up, then forward, making a counterclockwise circle, from your perspective. Properly swung, the houlihan loop should stay wide open, and your thumb should be down as the loop comes forward over your head.

Even though the houlihan travels at the same angle as a sidearm, it requires more arm action to give the loop the momentum it will need for long-distance catches. As you swing, extend your throwing arm well behind you as you bring the loop back, up, and forward again. Like a forward swing, the houlihan requires very little wrist action.

As an exercise, try raising your elbow and hand, flattening the angle of the houlihan so that it travels overhead. This is not a practical working shot, but switching from a traditional houlihan to this overhead houlihan and back will help you improve your control of the rope.

There is a widespread misconception, even among veteran ropers, that a houlihan should be preceded by only one swing. In reality, the houlihan can be thrown after any number of swings, but in a working situation, a single swing has the advantage of causing less of a commotion, decreasing the chances of spooking a group of cattle. ❧

1. The houlihan travels at an angle similar to that of the sidearm swing, but in the opposite direction. The roper's thumb should point downward as the houlihan travels overhead.

The Houlihan

2–4. Buck practices the houlihan, positioning himself on the roping dummy's right side, as if targeting a cow traveling from left to right.

5–7. The houlihan is essential to good outside roping, as you can throw long distances without disturbing the cattle as much as you might with another swing. This sequence shows how Buck holds his coils so they feed off well and illustrates the houlihan swing and throw.

5.

6.

7.

PRACTICING THE HOULIHAN: As outlined in the section on practicing forward swings, a roping dummy can be a good tool for early practice making houlihan catches.

Position the roping dummy in an open area, and step back ten to twelve feet.

Build your loop as explained earlier, and begin a houlihan swing. Make your throw, releasing the loop as your right arm is fully extended toward the dummy's head, and follow through with your right hand.

Work at developing a feel for the amount of energy you need in your swing for the loop to reach the dummy and not overshoot it. Also, try to achieve a sense of how your arm action affects the loop's speed, your control of the loop, and ultimately, the accuracy of your shot. Simply put, this requires firsthand practice, and lots of it.

When you have made your catch, use your right hand with your palm down to pull your slack straight back to your right hip. At this moment, check the position of your right hand; this is the point where you will need to dally your rope onto your saddle horn.

Next, retrieve your rope, recoil it, and after you have ensured your coils are organized and of even size, rebuild your loop for another houlihan swing.

At first, it is best if you make multiple houlihan swings before throwing your loop. As you build your skills with the houlihan, you can work toward making fewer swings until you are able to make a respectable houlihan throw with just one swing. When you reach that "one-swing" phase, you can add momentum to that single swing by first rocking the loop back and forth like a pendulum.

Once you are comfortable with the houlihan, bring the set of forward swings—sidearm, overhead, and overhand—back into your practice sessions on the sawhorse roping dummy.

The Backhand

Most casual observers of ranch ropers find it difficult to distinguish between the houlihan and the backhand swing. Both travel in the same direction—making a counterclockwise circle from the roper's perspective. But important differences in the techniques are at work, and in the hand and arm positioning required for each of these swings.

To develop your skills with the backhand swing, begin work on the ground rather than on horseback, and keep your feet square as if they were in your saddle's stirrups. Start by holding your loop in front of you with your palm down. Flip the loop backward and to your right, as if you were merely building a loop. Let the rope turn over itself as you flip it. With your right arm extended to your side, hold your position. Your palm should now be up.

Practice this maneuver until you are comfortable with the arm movement necessary to consistently position the rope to your right side with your palm up. When you are ready to proceed, repeat this movement, but continue your swing so that the loop makes one full overhead revolution. As you swing, rotate your fist so your palm remains up after the loop travels past your head.

Make just the one swing and then bring the loop to rest on the ground at your feet. Your honda should now be to the right of the loop.

Work on these two sequential movements, finishing with no more than a single swing. This will help you develop correct backhand technique, and help prevent you from reverting to a houlihan swing. Once you are proficient with a single backhand swing, advance to making two swings before bringing the loop to rest on the ground. Then, work on making multiple swings.

1–2. Begin the backhand swing by holding the loop in front of you, palm down. Flip the loop backward and to the right, ending the movement with your palm up.

The Backhand

3. As you make the backhand swing, rotate your fist so your palm remains up after the loop travels past your head and continues on your right.

4. Buck uses a backhand flank shot to make this heel catch. To make the catch, he rode his horse behind the cow on a path perpendicular to the cow's position. As Buck's stirrup became even with the cow's hind end, he sent the loop in front of her left hind leg. When the cow steps forward, her hind legs will be in the loop, and she will be heeled.

5. Buck prefers to "split the spoke" for the backhand flank shot. This allows him to maintain control of the spoke after he has released the loop.

6–8. Working on foot, Buck makes a practice flank shot using the backhand swing. He walks on a path perpendicular to the dummy's position, and as he comes even with the dummy's back end, sends the loop in front of its left hind leg. Buck maintains his hold on the spoke as he delivers the loop.

The Backhand

7.

8.

PRACTICING THE BACKHAND: Again, use a roping dummy in your initial practice sessions with the backhand swing. As described earlier, position the dummy in an open area. The first backhand shot learned is the backhand flank shot, a heel shot that is thrown while the roper is in motion, traveling behind the cow.

To practice this shot, position yourself so that the roping dummy is ten to twelve feet in front of you and facing to the right. Begin your backhand swing and walk forward on a path that will take you behind the dummy, a few feet from its back end.

As you pass behind the roping dummy, your backhand swing in motion, imagine yourself on horseback. As you ride past the dummy's back end, deliver the loop so that you send its tip in front of the dummy's left hind leg. Done correctly, this will position the open loop in front of both of the dummy's hind legs. If the dummy were a live cow, when she stepped forward with her hind legs, she would be caught in your loop. If you were on horseback, you would pull your slack, cue your horse to face up to the cow, and then dally onto your saddle horn.

After you have delivered your practice swing, retrieve your loop, recoil your rope, and once your coils are correctly organized, rebuild your loop to make another swing.

Once you are comfortable with the backhand swing, practice it in conjunction with the other basic swings—the sidearm, overhead, overhand, and houlihan. Place your roping dummy in the center of a corral or open area, and practice making catches from various positions with the dummy at the center of a circle with a radius of about twelve feet. ✺

9-11: Buck uses a backhand swing to make a flank shot during a doctoring session in his corral.

The Backhand

10.

11.

Ranch Roping

BASIC HEEL SHOTS: A ranch roper needs to be proficient in three other fundamental heel shots in order to be an effective hand: the trap shot, the flank shot, and the hip shot. We will take a look at each in the following sets of photos. ✤

1–3. The trap shot is thrown from an overhand swing with your horse just to the left and straight behind the cow. As with all trap-type shots, you hold the spoke and let the loop "drop out" and wrap around the cow's legs.

The Backhand

[33]

4–6. Buck uses an overhand swing to make a heel catch on this cow during some corral work.

The Backhand

7–9. The flank shot is taken from a sidearm swing, with the cow traveling from left to right, or with the cow stationary and the cowboy riding past.

10–12. The hip shot is the first heel shot you will learn with which you throw both the loop and the spoke. The object is to have the base (the part of the loop that is in your hand), rather than the tip, reach the target first, to catch the hip. The tip then follows through and wraps around the hind legs.

The Backhand

13–15. Buck puts the hip shot to work in the corral.

[37]

Ranch Roping

A WORD ON PRACTICE: To develop the feel and comfort level you will need to put your new tool—your rope—to work, you will have to spend a great deal of time with a rope in your hand. Carry a rope as often as possible, and rope any target you can't hurt. Roping is not a casual pastime; it requires dedication on an almost spiritual level and firm commitment to the premise that there is no such thing as too much practice.

TARGETS: A roping dummy gives a roper a more realistic mental picture of the positioning required for a given swing, without the risk of hurting yourself, your horse or any cattle. If you are a moderately skilled carpenter, you can build a roping dummy—again, a basic sawhorse fitted with a plastic steer head—relatively easily, using two-by-four lumber or, for that matter, scrap lumber. A plastic steer head can be purchased from any ranch supply store.

Once you have become proficient at roping the dummy from the ground, get on horseback and continue throwing practice shots on the roping dummy. As you might expect, from horseback, your perspective will be different. The positions you have become accustomed to on the ground will be replaced by a new set of angles, thanks to your elevated view from the saddle. ❦

1. A roping dummy, essentially a sawhorse fitted with a plastic steer head, makes a great practice target for beginning and experienced ropers alike. While roping on the ground, it is important to position yourself as if you were on horseback. Here, Buck has made a head catch and pulls his slack straight back with his right arm. He has brought his left arm forward, just as he would on horseback to get the coils out of the way so that he can dally. Here, he works with a traditional, hand-braided rawhide reata.

2. Once you have put in enough on-the-ground practice to consistently make catches on the roping dummy, it is time to get on horseback. From that point, it is best to saddle your horse for every practice session. Because of your height off the ground, your perspective as a roper changes as well as the angles with which you work. If you have progressed to the point where you are ready to rope from horseback, you might as well make practice sessions as realistic as possible. Here, Buck uses an overhead swing to make a heel catch on the roping dummy.

Buck Brannaman on
Ranch Roping's Heritage

There's a kinship among those of us interested in this kind of roping. We're also interested in this kind of horsemanship. We all might do things just a little bit differently, but we're looking for the endgame to be the same: your basic California vaquero-style horsemanship and roping.

—BUCK BRANNAMAN

There are different types of cowboys. The southwestern cowboy, the Texas cowboy, will rope hard and fast with the rope tied to the saddle horn, or with a shorter rope, or with rubber wrapped around the horn. In the vaquero style, ropers use longer ropes and some elaborate loops. Some of those loops might be fancy, but there is a practical application for all of them, in the right circumstances. Anyone who can get the job done and done well, though, whatever kind of cowboy he is, deserves respect.

The vaquero style of roping has always gone hand in hand with the vaquero style of horsemanship. Both are very practical in the way horses and livestock are handled, but they are both sources of pride for buckaroos, as they were for the old vaqueros.

We might all have different approaches to getting the job done, but we all want to get the job done in such a way that the horse is responsive, in a good frame of mind, and comfortable. If the horse is not comfortable with himself as he does this kind of work, he feels a lot of pressure and undergoes stress. He is troubled. When you are in the heat of battle with a cow, a troubled horse will betray you.

Someone I would consider a great roper also knows how to set up shots for other cowboys; he does not act like it is all about him. He considers the people he is roping with and makes them better ropers by setting up shots for them. All those elements enter into it—being a horseman, a cowboy, a hand, a team member.

Great ropers know how to be safe. They have an awareness, whether it is in the pasture or in the branding pen. They can plan ahead and look out not just for themselves, but for others, as well. Odds are that someone with that sort of awareness and consideration will also be considerate of his horse. And, anyone that considerate of his horse is apt to be a pretty good horse hand, as well.

The hands I have been around, enjoyed working with, and enjoyed watching, they are interested in being horsemen, not just in getting the job done and getting back to the bunkhouse. They are the whole package. They want to do the job right and do the best thing they can do for the horse, all with some style and some elegance. Having it done well—with elegance, style, skill, and precision—that means a lot to us. It all fits together. It is part of the same package, and it is what makes being a buckaroo appealing.

There is a basic human respect for other riders that is carried out to the pasture where the cattle are handled. It is a matter of being a gentleman, really. There is an etiquette to it. You wish everyone had that mindset when they were driving on the highway; things would be a hell of a lot different. It is just a matter of trying to be a good person.

When you are making a circle and you are moving a bunch of cows, you do not ride in front of somebody. That is an absolute show of disrespect. If you need to move your way around the rodear, you might ride behind another rider and reposition yourself, but once you have made your throw, you do not stand there and hog the area. You ride your horse out so another roper can come in.

We practice. We do not just go out and doctor cattle when they are sick or dying and hope like hell it works out. Odds are it would be a big wreck and *not* work out. We practice our skills as horsemen and as buckaroos so that when we need to do the job, we can step up and do it. We practice our game so that when it is game time, we can deliver.

I am not the only person who teaches ranch roping. Peter Campbell is a great roper, as are Dave and Gwynn Weaver, Joe Wolter, and Bryan Neubert. Dave and Gwynn have done a lot through the annual Californios Ranch Roping & Stock Horse Contest to preserve this style of roping. I am thrilled they are all doing it. They are great people, and great friends.

As far as reata-style roping goes, the Californios is like the Super Bowl, our U.S. Open. But even though it is a contest, it is really a bunch of friends who are there roping together,

As Buck watches, Joel prepares to make a throw into the rodear, which is gathered in the corner of a pasture.

rooting for one another, and on any given day, any person can win it.

Competition is really secondary to the fellowship that occurs when everyone gets together and enjoys roping and sharing their skills and what they have learned in their lives.

Those of us who enjoy roping share a lot back and forth. It is not like we are trying to hide our secrets from each other. We figure if you are good enough to copy what we are doing, then you should be doing it. This is not a competition for us; it is camaraderie.

There is an inner circle of people who are good ropers, and they are all there for the common cause: we love the tradition of this ranch-style roping and reata roping, and we want to do what we can to preserve it. The reverence for tradition trumps everything else.

Even after you have been roping for years, you will still see another guy that might swing a particular loop with a different angle, or position his horse differently, or have his own way of taking slack. There are lots of nuances to roping that go beyond throwing a fancy shot and so much technique involved in handling a rope that completely.

Ray Hunt used to houlihan those colts he worked with. I grew up around guys that would throw a houlihan with just one swing—a single swing and a throw. They might drop a coil or two to send the rope a longer distance, but the houlihan was really for necking calves. It was an unusual swing, kind of plain, with a small loop to it, but you admired anyone who could throw one. Ray was the first guy I ever saw who could *swing* a houlihan—make multiple swings, instead of just a single swing—and actually deliver it. I learned a lot from that, and it did not take me long to get on to it myself. Because of my experience as a trick roper, I could see someone throw a shot and, in a couple of days, I was on it. I could get a rope to do what I wanted it to do, if I could just see it done.

I watch other ropers as if it is my first time watching anyone rope. I am interested in becoming a better and better roper, so when I am around a penful of good ropers, nothing will keep me from learning something from them.

Chapter Two: The Mechanics of Dallying

To become a good roper, you dally as many times as you throw.
—BUCK BRANNAMAN

Dallying	43
Logging	52
Tracking	63

Dallying

When a roper catches a cow around the head or heels, he or she pulls slack out of the rope to tighten the loop, then dallies, making one to three wraps around the saddle horn with the slack that is still gripped in the roper's throwing hand. The horn's surface holds those wraps—the dallies—secure and in place. In this section, we cover the mechanics of dallying and safe ways to learn and practice this skill.

Your horsemanship skills and your horse's level of training must be up to par before you begin working on ranch-roping techniques from horseback. Before you begin making swings from the back of a horse, take stock of your horse's readiness to work with you. Is your horse still learning the fundamentals of basic handling, or is it a seasoned and steady horse you can rely upon as you add to your skills and your horse's? An experienced horse is a novice roper's greatest asset.

A solid, reliable roping horse should have a quiet disposition, respond readily to cues, move laterally with its hind and front quarters, stop on cue, back readily, rein one-handed, work quietly while a rope swings overhead, be familiar and comfortable with a pull on the saddle horn, and be familiar with the feel of a rope against its body, around its feet, and under its tail.

SADDLES: Dallying requires a saddle built for the job. A well-made saddle with a laminated wood horn will handle the pressure created when a dallied rope goes taut with the weight of a cow on the other end. A saddle horn with plenty of surface area around its neck offers a more secure grip on the rope, as more of the rope will be in contact with the horn.

Wrapping the horn in mule hide protects the horn surface from wear and provides a grip for dallies. But unlike a rubber-wrapped horn, a

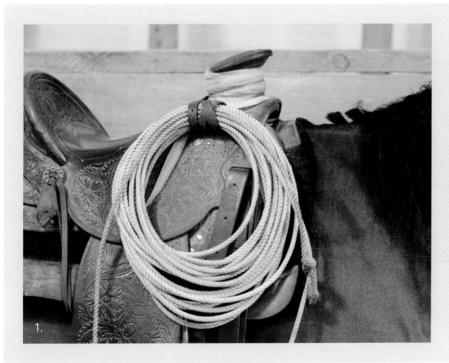

1. This traditional Wade saddle, owned by Arizona cowboy Joel Eliot, features the stout, heavy-duty horn typical of the style. The Wade has its roots in the Great Basin, but thanks in large part to the national, even international, interest in the horsemanship traditions of that particular region, Wade saddles have become mainstream far outside the boundaries of buckaroo country.

mule-hide wrap still allows a roper to "let rope run" over the horn and feed slack back toward the cow should a working situation require more distance between the horse and cow without letting the cow go altogether. Plus, just like the drag on a fishing reel, feeding slack over the horn allows the rider to taper the pressure placed on the cow, making it easier for the horse to work.

A saddle should sit as close as possible to the horse's withers without actually touching the withers, otherwise the saddle will make the horse sore. The combination of a breast collar and a snug back cinch provide added stability, helping keep the saddle in place as the weight of a cow pulls against the saddle from the front, back, or side. ❦

2. This saddle belongs to Buck and was made by Idaho saddle maker Dale Harwood. A Wade saddle horn's considerable surface area puts more of the rope in contact with the horn when a cowboy dallies. Mule hide on the horn allows enough grip for the cowboy to confidently keep his dallies in place, but unlike a rubber-wrapped horn, still allows a rope to be run over the horn's surface, meaning the cowboy can feed rope back toward the cow while keeping the dallies in place around the horn.

3. A snug back cinch helps stabilize a saddle. The back cinch should not be so tight as to inhibit the horse's air but should not be loose enough to have no effect or to create a safety hazard. A back cinch that is adjusted too loosely can get hung up on gates, fences, and brush, and can even trap a horse's hind foot. A set of hobbles hang from Buck's saddle, just behind his left hand, and Buck's lead rope, part of the McCarty, is draped over his arm.

Dallying

HOLDING REINS AND ROPE: Holding a horse's reins and the coils of a rope can make for an awkward handful. Buck recommends holding the reins between the left thumb and forefinger, then adding the coils to the left hand, resting them in the palm and keeping them in place with the other three fingers.

"With a McCarty [a set of closed reins with an attached lead rope], I will cheat my right rein a little shorter than my left," Buck says. "That way, if anything goes wrong, I can turn my horse to the right, even though my right hand is not on the reins. It is important the horse be maneuverable. When you ride a horse in a snaffle bit and use just one hand, things can really go wrong."

Keep track of how you hold the coils and reins in your left hand. Remember that if you shift the position of the coils, you will also shift your reins and unintentionally cue your horse to move.

Safety Point: When I step off my horse, I lay my coils on the ground instead of leaving them on the saddle horn," Buck says. "If I know where my coils are, I do not have to worry about getting tangled in them."

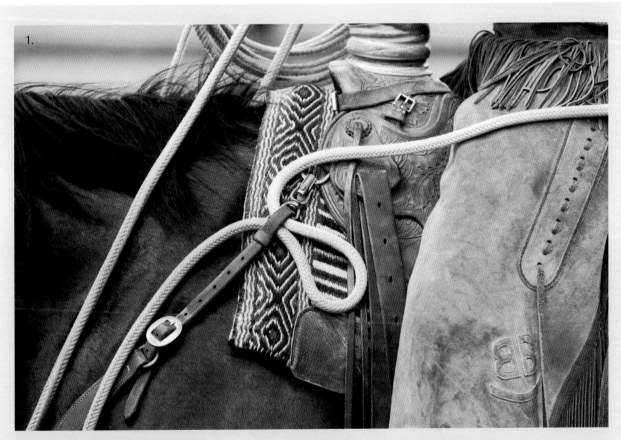

1. Buck has tucked the lead rope of his McCarty into the left strap of his saddle's breast collar. Note that he has doubled over the lead rope rather than running it through the opening in the strap.

2–3. Buck holds his McCarty reins between his left thumb and forefinger, and rests his coils in his left palm. To ensure he can easily turn his horse to the right with this hand position, he holds the right rein a little shorter than the left.

Dallying

Buck makes one exception to that rule: "With a reata, I leave it on the horn, but coiled, so that a horse will not step on it and 'strand' the reata." ❧

DALLYING MECHANICS: Once a cow has been roped and the roper has pulled the slack, it is time to dally. The most fundamental rule of dallying, according to Buck, is that ropers should never take their eyes off the cow to look at the saddle horn.

"Looking down at the horn is one of the most dangerous things you can do in roping because you are not keeping track of where the cow is," he explains. "It is amazing what can happen the instant you look away from the cow. She can wrap you and your horse in the rope, and it can become a disaster."

To dally, a roper first moves the left hand up the horse's mane. This moves the reins and coils out of the way, creates room to dally with the right hand, and keeps the roper from unintentionally turning a horse left while dallying.

With the right thumb up, the roper moves the right hand to the left, bringing the slack across the front of the saddle horn, and wraps

1. To dally, a roper brings the reins and coils forward with the left hand.

Ranch Roping

2–3. With the right hand, the roper brings the slack across the front of the horn, wraps the rope around the horn's base, and brings the hand back to the right.

Dallying

4–5. Buck finishes his first wrap around the horn. The number of dallies the roper makes depends on the size of the animal being roped. A small calf might require one dally; an adult cow might require three.

the rope around the base of the horn, bringing the hand back to the right.

The number of wraps the roper makes depends on the size of the animal being roped. A small calf might require only one dally; a yearling or adult cow might require two or three.

"You judge that on how well you think you can close your hand around the rope and stop the cow," Buck says.

An experienced roper can dally while riding away from the cow that has just been roped; this puts less stress on the horse's back. But green ropers should dally while facing the cow and backing away, a less-complicated task set than riding and reining a horse while simultaneously handling the dallies and keeping track of the cow.

"Until a person's roped and taken down a couple thousand head, he should face up to the cow to dally so he does not get wrapped up in the rope," Buck says. "Do not get in a rush to dally while riding off."

Safety Point: If you have missed your dallies, or are late getting to the horn, ride to the cow without stepping your horse over the rope. This buys you some time to regroup and dally again.

6. Having made his catch, Joel pushes his coils and reins forward along the horse's mane with his left hand and pulls his slack with his right hand.

Dallying

7. Joel has begun dallying after making his catch. With his left hand, he has moved his reins and coils away from the saddle horn, giving himself room to dally. With his right hand, he has brought his rope across the front of the saddle horn and taken two wraps. Joel has his dallies crossed to lock down the rope on the horn. If he wanted to run rope, he would remove half a dally and step his horse back. The number of dallies a cowboy makes depends on the working situation; a small calf might require just a single dally, while a bull might require three. The saddle horn shown is wrapped in mule hide, which allows a cowboy to "let rope slip," feeding slack back toward the cow should a situation call for it. Joel's horse carries a two-rein outfit, which includes a spade bit, with romal reins, and a bosal, with a McCarty.

Logging

When it comes to getting a novice roper, or for that matter a green horse, accustomed to working a taut rope, dallying, handling pressure on the saddle horn, dragging an object, and maneuvering with weight on the end of the rope, there might be no better training aid than a heavy log with a notch at one end. The log can serve as an ideal tool in developing the skills of the horse and rider, simulating for both the feel of a cow on the end of the rope while still giving the rider a high degree of control.

By securing the loop around one end of a log, riders can practice rope- and horse-handling in a controlled environment and develop their skills and those of their horses before they begin roping cattle.

Logging should be practiced in a safe area, ideally in a round pen.

Logging helps teach a horse to handle weight and pressure on the end of a rope, mimicking the feel of a cow, but giving a rider a higher degree of safety should the horse become uncomfortable or afraid. To be useful as a training tool, a log should have a notch cut into it, about two feet from one end, to help secure the loop in place as the log is worked. The log can be dragged with the horse moving forward or backing away, or the horse can drag the log in a circle.

Logging

GETTING STARTED: To begin, dismount and position your loop snugly around one end of a log weighing about one hundred pounds, about two feet from the log's end. A notch cut into the log where the rope will rest can help keep the loop in position and prevent it from slipping off as you work.

"When I put the rope on the log, I have the coils with me," Buck says. "I do not leave the coils of my rope on the saddle horn. Otherwise, if the horse got scared, he could take off dragging the log."

For safety's sake, before you remount, be sure you are prepared to immediately drop the coils to the ground in the event your horse panics at this unfamiliar scenario, or if another potentially dangerous situation develops.

Once you are on horseback, with the horse positioned about ten to twelve feet from the end of the log, begin backing your horse. As the rope goes taut, dally twice around the horn and continue backing the horse.

As you back your horse, keep it positioned so it faces the log. This is an important habit to develop. Once a cow is roped, your horse should stay square with the cow, facing her so she can't pull on the rope from the horse's left or right side.

The first time you try logging, your horse might be cautious and tentative, taking only a step or two with the weight of the log on the end of the rope. Be careful not to push your horse past its comfort level. You might drag a log many times before your horse settles and becomes fully at ease with the idea.

1. Buck has secured his loop around the log. For safety's sake, he has brought his coils with him rather than leaving them on the saddle horn.

[53]

"When I first start to back my horse, he might get worried," Buck says, "so I have to be ready to pop the dallies off my horn by lifting straight up with the rope. It is a much more efficient way to get the rope off the horn than unwrapping the dallies.

"When you suspect you are about to get into a dangerous situation, the order of events to get yourself out of trouble is, get the rope off the horn, drop your coils to the ground, then get hold of both reins to get control of your horse," Buck says.

"Remember that the rope is not tied off to the horn, so if anything goes wrong—if the horse gets scared, for instance—you can pop the dallies off the horn and, if necessary, drop the coils to the ground."

Drag the log a few feet, and then as you continue backing your horse, begin running rope. Let the rope slip around the horn, feeding slack toward the log. As needed, feed more rope to the dallies by dropping one coil at a time from your left hand.

"If my rope is on a calf and I think it might be pulling too tightly, hurting the calf, I can slip rope to lessen the pressure on the calf's legs without letting him go," Buck says. "That is important to me, especially if I am the one who owns the cattle. And the only way you can run rope is to have those coils straight and organized in your left hand."

By running rope, you lessen the pressure your horse feels from the weight of the object on the end of the rope. To your horse, the pull of

2. If a horse becomes nervous or afraid, or if an otherwise dangerous situation begins to develop, the dallies can be removed instantly by lifting straight up with the right hand, "popping" the dallies free of the saddle horn.

the log—or of a cow—will be a fraction of its full weight. To stop the rope from running around the horn, you close your right hand around the rope and stop it feeding toward the horn. To resume running rope, open your grip, keeping your hand well away from the saddle horn.

Next, stop your horse, pop your dallies by lifting straight up with your right hand, recoil your slack, and ride toward the log again, keeping the rope taut by recoiling slack as you ride forward.

"I teach my horse to walk straight alongside the rope so I can recoil the rope just as if I am 'shortening up' on a cow," Buck says. "You will coil your rope at the same pace you walk your horse. If you step your horse faster than you are coiling, he might step over the top of the rope and risk getting tangled."

To build your early confidence handling a rope, perform this exercise repeatedly—dallying, dragging the log a few feet, then running rope before stopping, popping the dallies, and riding forward to recoil.

In subsequent sessions, you might opt to place the loop around the log while you are still on horseback. To do this, you ride to the left side of the log, with the notched end below your right stirrup, and hook the loop in the log's notch. Draw up your slack by raising your right hand, ride forward, and use your right foot to ask your horse to roll its hindquarters clockwise so that the horse turns to face the log. Slowly back your horse to tighten your rope.

"Practice good horsemanship while you are logging," Buck says. "For instance, it is important

3. Buck runs rope over the saddle horn as he drags this log. His coils are straight and organized in his left hand.

that a horse learns to work with his chin down and in. If his chin is tucked, he is able to push back. It is a balance issue, and part of proper form and posture for a horse when he is stopping, slowing down, or backing up. But if a horse throws his head when you pick up on the reins, he can get his head over or under a rein."

SECOND STAGE: Once the horse is comfortable dragging the log backward, the next step is dragging the log in a circle, ideally inside an arena or a fifty- to sixty-foot-diameter round pen, which makes for a safe, controlled environment to work on a new skill.

With your loop around the end of the log, position your horse perpendicular to the log's end, and dally your rope around the saddle horn. Ride in a clockwise circle, keeping the rope angled away from your horse's body while making the log pivot like the second hand of a wristwatch. This gives your horse a chance to see the log move in a different way without the tightened rope being close to its hindquarters and making it afraid. This is an especially important point when you are riding a greener horse, as the unfamiliar sensation of a rope against its hindquarters might spook it.

4–5. Buck has dallied on and pulls the log in a circle with the log well to the side to get the horse used to seeing it. Buck then faces up and backs his horse.

Logging

Next, use your right leg to cue your horse to roll its hindquarters left and face up to the log. Then back your horse to finish the maneuver.

"If you do this perfectly, the tension never leaves your rope," Buck says. "Going forward, facing up, backing—the rope should stay tight. When you have heeled a cow, if you put slack in the rope, you will lose her feet."

As you repeat the pivoting exercise, gradually increase the size of the circle you ride. Soon you will no longer be pivoting the log but dragging it in a circle, with the spoke of the rope within about six feet of the horse's hip. Again, face up by using your right foot to ask your horse to roll its hindquarters left, keeping the rope tight, and then back your horse.

Safety point: If the horse spooks and whirls left, it could wrap you in the rope. Should this happen, pop your dallies off the saddle horn (again, by lifting straight up with your right hand), take hold of your right rein with your right hand, and face up to the log. As you get more confident, you will be able to get the rope closer to the horse's hind end and drag the log around the arena.

"When I roll a horse's hindquarters and face him up to the log," Buck says, "I always let the rope slip in that moment when I am sideways to

5.

the log. When a horse is sideways to a real cow, that pull from the side on the dallied rope can turn a saddle sideways and stress the horse's back. Slipping rope makes a two-hundred-fifty-pound calf feel like one hundred pounds to the horse. Then, when I have faced up to the cow, I stop slipping rope and take the entire weight of the animal again."

Keep your expectations realistic when you introduce this maneuver to your horse, and allow it to build its comfort level over time and over multiple sessions.

6. Pay close attention to your rope's position as you work, especially when you are aboard a green horse. The unfamiliar feeling of a rope against a horse's body can trigger panic in a young horse still getting accustomed to the routines of roping. Early on, keep the rope angled away from your horse. Gradually, as you introduce new maneuvers, your horse will become more comfortable in its role, and it will be less concerned with the rope's positioning. Buck prefers to practice holding a tight rope, with his horse facing away.

Logging

RUNNING THE ROPE LENGTH: Many beginning ropers do not have easy access to cattle, and they assume they can't become adept at dallying. This no-cattle exercise will help a beginning roper become quick and efficient at dallying.

You will need a lightweight log, one weighing around seventy-five pounds, with a notch cut into it about two feet from one end.

Begin with the loop around the end of the log and secured in the notch, and the rope undallied. Ride off, and as the rope tightens, dally on, then let the rope run over the saddle horn as your horse travels several feet.

Then pop the dallies from the saddle horn by lifting straight up with your right hand. Dally on again, let the rope run over the horn for several of your horse's strides, and pop the dallies once more. Continue that pattern until you run out of rope.

"This makes you dally a tight rope," Buck says, "and teaches you to let a rope slide through your hand, which helps you avoid a rope burn. When you are running rope, let the rope go through your hand easily and let the horn take the friction."

As a horsemanship exercise, you might ask your horse to back the length of the rope as you recoil. This exercise can be done only if your horse is especially comfortable dragging a log, first by backing, then by going forward. ✹

1. Buck runs the rope length. He rides off with the loop around the log and the rope undallied.

2–3. Next, he dallies on, lets the rope run over the horn for several strides, then pops the dallies, continuing that pattern until he reaches the rope's end.

Logging

HEEL-SHOT SIMULATION: With the loop secured around the log and the rope undallied, ride away from the log to remove slack from the rope. As the rope begins to tighten, push your left hand forward, up the horse's mane, and dally with your right hand. Let rope slip around the saddle horn so your horse can gradually ease into the weight of the log.

"At some point [during the above exercise], I might stop my horse and ask him to hold the rope tight, as if I am holding down a cow," Buck says. "I need him to hold the rope tight and not fidget or move his hindquarters. You need the ability to control the horse's hindquarters with your legs, and keep yourself out of trouble."

After the horse stands for a minute or two, Buck adds, he might ask the horse to move forward, slip rope around the saddle horn, then stop and ask the horse to hold the load again. This exercise can be continued until the full length of the rope has been run.

HORSEMANSHIP NOTE: "One classic mistake ropers make is dallying when the horse is going too fast," Buck says. "The roper thinks he will use the cow to stop the horse, that he will just stack enough dallies on so that when things come tight, everything will stop. He forgets he has reins in his left hand, and that all he needs to do is pick up those reins to slow the horse."

1. Buck simulates the follow-up to a successful heel shot. He rides away from the log to remove slack from the rope.

2–3. As the rope tightens, Buck dallies. He then runs rope around the saddle horn, gradually letting the horse ease into the full weight of the log.

Tracking

If your horse lacks experience on cattle, tracking will help build its confidence and help it learn to "hook on" to a cow.

"Tracking helps a horse learn to set you up for a great head shot from straight behind the cow," Buck explains. "It also teaches a horse to drive the cow. You might come upon a cow that needs to be roped, but she is not in a convenient place. Your horse needs to be able to drive her to a spot where she can be roped."

Begin in an arena or similar enclosed area that offers plenty of working room. Ideally, you should use an arena with a high fence and ground that is not slick.

Build a loop and let it drape from your right hand. At first, you will track a cow without swinging the rope.

Start by riding toward a cow and simply following her as she moves around the arena. Try to keep your horse positioned just behind the cow's left hip, a common spot to throw a head loop from.

"Some of this is done at a walk, some at a lope, and you will do a lot at the trot," Buck says, "You do not want to get in a huge hurry. This is not about speed, but about teaching your horse position."

As the cow increases her speed, change gaits to keep up, but do not try to wear her out. Just close the distance between your horse and the cow so that she does not get away from you.

Next, drive the cow down one side of the arena, positioning your horse behind and to the inside of the cow to keep her against the fence. After you have made a full circle around

1. Tracking helps teach a horse proper positioning. A rider can begin by simply following the cow, progressing toward driving her down one side of an arena or other enclosure, keeping her against the fence.

the arena, reverse directions and take the cow around the opposite way. As you ride, keep your horse behind and to the side of the cow. Work on keeping control of the cow and not letting her drift into the middle of the arena; instead, keep her on the arena perimeter.

At first, the cow might resist being pushed in a specific direction. Be persistent in using your horse to control her and keep her along the arena perimeter.

As the cow approaches a corner, give her plenty of room to turn the corner so that she does not come to a stop where the fence makes a right angle. As the corner nears, point your horse toward the cow's outside hip (the hip closest to the fence). Give the cow room to make the turn; once she is through the corner, again position your horse behind and to the inside of the cow.

"Do not let the cow turn back on the fence," Buck cautions. "If she does, block her."

Continue tracking the cow around the arena perimeter in each direction until she begins to tire. Your horse should begin reacting to the cow's movements and developing a sense of how to move *with* the cow rather than independently of her.

Next, work on tracking the cow while swinging your rope. This will help accustom you to swinging while your horse is on the move and reacting to the cow's changes in direction.

Work up to making a throw at the cow, but instead of trying to make a real catch, simply make contact with the cow by throwing the loop over the cow's hips. This helps accustom your horse to the sight and sound of a rope being thrown past its head. Practice different swings, and shift the angles of your swings, again to familiarize your horse with the motion of the rope.

Once your horse seems comfortable tracking, and you are able to maintain control of the

2. Gradually work toward throwing a loop at the cow. To help accustom a horse to the sight and sound of the rope in motion, at first, you might let the rope simply make contact with the cow rather than try for an actual catch.

Tracking

3. A breakaway honda, shown here on the loop, allows for quick retrieval of the rope during tracking sessions.

4. A knotted rope's fixed loop size keeps the loop from tightening, allowing the cow a quick escape, and gives your horse a more realistic pull, not unlike what it will experience in a working situation.

cow, you can attempt to make head catches.

"At first, you might want to track the cow and just swing the rope, letting the horse get used to the rope being swung while he is behind the cow," Buck says. "When it looks like I am getting a nice long run down an arena or small pasture, I will let my horse 'track up'—get positioned behind and to the left of the cow—so I can take a head shot."

Begin by using a rope with a breakaway honda so that when you pull your slack and make your dally, the rope will break free of the cow's head. This way, you can practice head catches alone without having to worry about retrieving your rope.

Build a loop and begin an overhead swing while following the cow. When the opportunity presents itself, throw a head loop. When you catch, pull your slack and dally. When the loop breaks free of the cow, recoil your rope, build a new loop, and begin again, tracking the cow along the arena perimeter.

"Once I rope the cow, I want the horse to follow, not stop," Buck says. "I might have a cow that needs a calf pulled. I might need to take her to the barn on the end of a rope and need my horse to continue following, pushing her."

Once you are proficient with the breakaway rope, switch to a knotted rope, one with a fixed loop size. The loop will not tighten, so the cow will quickly escape, but a knotted rope will give your horse a more realistic pull, not unlike what it will experience in a real working situation in which you use a rope with a traditional honda. When the cow escapes the open, fixed-size loop, recoil and begin again. ✺

5.

5–8. Buck brings the cow around the end of the arena where he sets himself up for a successful head shot using a rope with a breakaway honda.

6.

7.

8.

Ranch Roping's Horsemanship Challenges

With roping, it's easy to get glazed over. You think about roping so much, you abandon the horsemanship you've studied all along. Then the more you rope, the worse your horse gets. You see folks that might be capable at roping, but they fall short because their horsemanship is so damn poor, they can only progress so far.

—BUCK BRANNAMAN

Getting the horse to tuck his head and have a soft feel, teaching him to drop his chin while you are holding your hands high, is critical to roping. Ordinarily, you would have your hands lower, in a more ideal position for riding, but to rope you have to hold your hands high and have the horse tuck his head with your hands in that higher position.

If you are making a forward swing from horseback, typically you will be going left with your horse. In harmony with your hands, you will put your left leg back and right leg forward to help guide your horse. To take my horse to the left, I ride with my left palm down and use the left rein to tip my horse's nose to the left, when using a snaffle bit or hackamore. If I had to take my horse to the right while making a forward swing, I would bring my swing back behind my horse's hind end so I would not hit him in the face with my rope.

Swing a nice, big loop so you are less apt to hit your horse. With a smaller loop, the entire action of the rope is confined to the same space occupied by the horse. And the momentum of a bigger loop will help you keep it traveling correctly.

When you are riding a young horse that has never had a rope swung on him, it is best to go left with your forward swings and to make houlihan or backhand swings when going to the right. That way the energy of the rope and the horse is going in the same direction. That can be important on a young horse. Gradually, he will get used to you swinging forward while going to the right.

I practice having my horse drop his chin and get soft while I am swinging my rope. I like the

A polished ranch-roping horse tucks its head and chin while the roper works with his hands in an elevated position.

horse to give to the snaffle, just as he should do when I do not have my rope down. I also practice stopping my horse and backing him with his chin down, all while I am swinging a rope from horseback.

Before I rope, I always make sure that while I hold the reins in one hand I can step my horse's hindquarters to the left or to the right, step his front end to the left or right, and side pass him left or right using my legs.

When you first build a loop aboard a green horse, or a new horse, start off by swinging the rope easily, just rocking the loop forward and back without actually bringing it overhead. Then stop and gauge how well your horse is handling it. If he is doing all right, you might make a single sidearm swing, then stop the motion of your rope. If your horse can handle that, you might rock the loop forward and back again, then take a couple of swings. With a young horse, or one you are not familiar with, this is when you might find out that he is not really ready to have a rope swung on him. The last thing you need is to rope a cow when your horse is troubled.

Once your horse can handle a sidearm swing, change the angle of the swing so the loop travels flat overhead, then angles forward so it travels in front of the horse's head, and then angles over your left shoulder so it travels in front of the horse's left eye. Then, get him used to the houlihan swing; coming from a different angle could concern the horse.

When you are first learning to swing your rope, and you stop your swing, the tip of the loop might follow through and slap the horse or wrap around his legs. You have to learn to take the energy out of your loop so that when you stop your swing, the loop lands beside you. When I swing my rope, it is as if my hand is the center of a hub. When I want to stop my swing, I move my hand straight up to the outside of the circle, and I will let all the energy go out of the rope. That way, it stops beside my horse rather than slapping him in the face.

Once you are roping on horseback, it is important you saddle your horse and mount up for dummy-roping sessions rather than rope on the ground. The angle your rope travels will be different when you are on the back of a horse. You are six feet off the ground, so your throw will be different than it would be if you were afoot. It is worth five minutes of your time to saddle up and rope the dummy from horseback. It also gives you a chance to work on positioning your horse, moving his hind or front end so you are perfectly situated to take a shot. That helps your horse get particular about where he will put you when you take a certain shot.

You have got to work on a loose rein. This is something people can spend a lot of time working on, and it is a key skill that does not require cattle to develop.

There is going to come a time when you have a rope get to an unfortunate place. Your horse needs to be able to tolerate having that rope end up around a foot, for example, and be comfortable with the rope being there. It will happen. Either you will put a rope there, or someone you are roping with will help you get a rope there.

A ranch horse must be able to tolerate a rope winding up in "an unfortunate place."

Ranch Roping's Horsemanship Challenges

Chapter Three: The Cowboy and the Herd

Being a cowboy is a craft that really takes a lifetime to learn. It's not an easy thing. It's a lifetime study to be a good hand, to be a good cowboy. It's a shame self-proclaimed intellectuals use the term as a dirty word when it takes less time to become a doctor than it does to become a skilled vaquero.

—BUCK BRANNAMAN

The Rodear	73
Working Scenarios	75
Three-Man Doctoring	90
Two-Man Doctoring	98

The Rodear

When cattle are worked in open country outside the confines of a corral, the *rodear* is where the action takes place. Spanish for "to encircle," *rodear* can be synonymous with a roundup, but in general the term refers to the bunching up of cattle in an open area and to the collected group of cattle itself.

Cowboys, or buckaroos, split up to ride assigned areas, gather small groups of cattle, and push them toward a predetermined location. A herd randomly scattered over a wide area gradually becomes several tighter-knit groups of cattle converging into one mass. This gives the term *rodear* the connotation of a rendezvous, a regrouping of cowboys as they bring in stock to add to the larger herd that is building.

To hold the cattle together, horseback cowboys surround the rodear at strategic points, perhaps holding the herd in the corner of a pasture when convenient. Generally, the cowboys position their horses around the perimeter of the herd, equidistant and facing the cattle. Riders need enough stock-handling experience to know how much pressure to apply and how closely to position their horses to the cattle. When riders are too close, the herd can't settle and will be chaotic and hard to handle. When riders are too far away, there might not be enough pressure on the herd boundaries to keep the rodear intact.

If riders have the benefit of a fence corner forming boundaries on two sides of the rodear,

Sometimes the rodear might be holding up in a corner. Here, Buck has roped a cow on the edge and eases her out so a bad eye can be doctored.

Kevin has roped a cow outside the rodear, which stands gathered in the pasture corner behind him.

they might maintain more distance from the herd than they would in open country for fear of pressuring the cattle so much they decide to test the fence. Even a stout barbed-wire fence is never a match for a cow that decides she is going through it, much less for an entire herd of like-minded cattle.

If cattle are being held in open country, the situation is a little more precarious. Cowboys have to pay attention to keep the herd together without creating a tight boundary that could disturb the herd and trigger a runoff or mini-stampede. Quiet control is the order of the day.

A rodear can be an opportunity to let cattle regroup, allow calves and cows to pair up, give the herd a chance to rest after being trailed some distance, or pause to create a staging area in advance of taking the herd toward corrals, across a road, or into some other situation with potentially tricky logistics.

In some cases, the rodear allows for sorting or separating cattle by category—for instance, in the case of a mixed herd, or a scenario in which all the cattle have been gathered but only a portion will be worked that day. One cowboy might ride into the rodear and cut cattle to a strategically selected spot. Riders positioned at that spot along the rodear's perimeter allow the cattle to quietly slip through the rodear boundary.

And, of course, a rodear is used for open-country brandings in which there are no corrals, only a perimeter guarded by mounted cowboys. A fire is built, irons are heated, and as the perimeter cowboys keep watch, a rider quietly enters the rodear to sort out an animal to be roped and taken to the fire.

"Working in a rodear, you do not want to rope a calf inside the herd and have the cattle so disturbed they will mash through the fence," Buck says. "With cows and calves, you go to the edge of the rodear, neck the calf, and pull him toward the fire for your heeler to come in and catch him *at* the fire."

When holding yearlings in a rodear, it is best to move the animal to be roped to the outer edge so it can be caught without scattering the rodear.

Working Scenarios

1. The cow travels from the roper's left to the roper's right; a head shot is needed.

Potential throw: sidearm head catch

Build your loop and make your sidearm throw, releasing when your arm is fully extended. Make your throw as the cow passes your horse's head, and remember to follow through with your right hand.

Watch your loop as it settles over the cow's head. If you have thrown your sidearm loop correctly, on the far side of the cow, a figure eight should form in the loop as it travels through the air, even if only for a brief moment. The figure eight helps take slack out of the loop as it settles over the cow's head. Without the figure eight, the loop would be so large that the cow could easily step through it, leaving her roped around the middle. When a loop is behind a cow's front legs and around her middle, she is nearly impossible for the roper to control.

When you have made your catch, pull your slack and begin to dally. The cow will continue traveling to your right, so use your right foot to ask your horse to pivot its hindquarters to the left and face the cow. This ensures the cow will not pull against the horse's side.

Scenario 1. Buck uses a sidearm swing to make a head catch on this cow moving left to right.

Ranch Roping

2. **A cow travels across the center of the pen, from the roper's left to the roper's right; a heel shot is needed.**

Potential throw: sidearm flank shot
Begin with an open sidearm swing, and cue your horse forward. As you pass behind the cow, about six feet from its back end, with your horse perpendicular to it and your right foot even with its tail, make your throw.

"When the tip of your loop can reach the cow," Buck says, "you are in the throw."

Send the loop in front of the cow's right flank so it lands in front of both hind legs. If the loop is correctly placed, one-third should be in front of the cow's hind legs, one-third should emerge from in front of the right leg, and one-third should emerge from in front of the left leg. The loop should be open vertically, with the top of the loop even with the cow's flank.

Let your horse take a couple of steps past the cow. When she steps into the loop, pull your slack, roll your horse's hindquarters left so your horse can face up, then dally and back your horse.

Safety Point: With such heel shots, watch for the loop to arc around both of the cow's hind legs so that the tip extends behind her left hind leg. Do not let your horse step into the open loop as you ride past the cow's hind end.

"Be sure that when you ride past the tail of the cow, you are about six feet behind her," Buck advises. "That way, when your rope wraps around the cow's hind legs, the tip of your loop will not come around and catch the front leg of your horse. I have seen loops wrap around that way and catch riders by the right stirrup. It can be a disaster."

Scenario 2. Buck makes a sidearm flank shot during some corral work. The sidearm flank shot travels in front of the cow's right hind leg. Buck makes his throw as his right foot comes even with the cow's tail.

[76]

Working Scenarios

3. **The cow travels from the roper's right to the roper's left; a head shot is needed.**

   Potential throw: scoop loop
   Begin a sidearm swing, then adjust your swing so the tip of the loop angles inward as the rope travels past your foot, and angles outward as it travels overhead. With this swing, a catch can be made if the horse is perpendicular to the cow, if the horse is nearly facing the cow, or at any position between those two extremes.

Scenario 3. Buck makes a head catch using a scoop loop.

Working Scenarios

4. The cow travels from the roper's left to the roper's right; a heel catch is needed.

Potential throw: left-to-right hip shot
Position your horse so that it is standing behind and to the right of the cow, with your horse's front feet at a forty-five-degree angle to the cow.

"You will learn to throw this shot from many different angles," Buck says, "but that is the best position in which to start."

Begin a sidearm swing and throw your loop at the cow's hips—not her legs—by pushing the loop with your palm.

"Throw the loop over the cow's tail, so it hangs on her hip," Buck explains. "It is as if you are throwing the loop at a flat wall, and you want all of the loop to touch the wall at the same time. If the tip of the loop gets there first, you will not get the loop hung on the cow's hip."

Delivered correctly, the base of the loop will drape over the cow's hips and hang down her side, in front of her rear legs. In tall grass or a muddy corral, this placement holds up the trap as the cow steps into the loop.

"From here, coil up your rope and walk toward the cow," Buck explains. "You will walk on past the left hip of the cow, roll your horse's hindquarters to face up, back away to close your loop, and then you will pull your slack and dally."

Scenario 4A–E. A hip shot can be thrown from many angles. The resulting throw, though, should leave the loop draped over the cow's hips and hanging down her side, creating a trap the cow will step into.

4C.

4D.

4E.

Scenario 4F. Buck has headed the cow, and Joel makes a backhand hip shot, draping the loop over the cow's hips so that it hangs down her side, setting a trap in front of her rear legs. This shot enters the opposite side of the standard hip shot shown in the previous sequence.

Scenario 4G. Joel seals the deal, squaring his horse up to the cow and pulling his slack so that the loop begins to tighten around the cow's rear legs. As the cow continues to step forward, Joel will maintain pressure on the loop, which will then settle and tighten around the cow's legs.

Scenario 4H. Here, we see Roland Moore's hip shot as it approaches its target.

5. **The roper's horse is behind the cow, slightly off her left hip; a head shot is needed.**

Potential throw: flat overhead swing
Begin your forward swing, flattening the angle of your loop so it travels overhead. Stand slightly in your stirrups, and with your arm fully extended, make your throw, following through with your right hand after the release.

As the rope settles over the cow's head, watch for a figure eight to form in the loop. Again, this helps take slack out of the loop, ensuring the loop will not be open enough for the cow to step through it and escape before the roper can tighten the loop.

The angle of an overhead swing can be adjusted for either a head catch or horn catch. If you are trying to rope a cow around the horns, the forward angle of the loop should be flat. A head catch requires the tip of the loop to be lower, passing right in front of your horse's head as you swing.

Scenario 5. Buck demonstrates "tracking up" on a heifer's outside.

Working Scenarios

6. **The cow travels from the roper's right to the roper's left; a head shot is needed.**

Potential throw: an overhand swing, with the tip of the loop angled slightly downward, over the roper's left shoulder

Begin your overhead swing, and adjust the angle of the swing so that it travels over your left shoulder.

Time your throw so that you release the loop just before the cow travels in front of your horse. Make your catch, pull your slack, and dally, moving your horse's hindquarters as needed to stay faced up to the cow.

Scenario 6. As Buck's loop settles over the cow's head, it begins to form a figure eight. This takes slack out of the loop, helping ensure the loop will not be large enough for the cow to step completely through it and escape.

7. The roper's horse is behind and to the left of the cow; a heel shot is needed.

Potential throw: an overhand heel trap, with the tip of the loop angled over the roper's left shoulder

As you make your overhead swing, with the loop angled as described above, stand slightly in the stirrups.

"The tip of the loop will pass in front of the horse's left eye," Buck says.

Hold on to the loop as you swing its tip downward so it travels in front of the cow's right hind leg, releasing only as the rope makes contact with the leg. Hold on to the spoke.

"When you deliver, open up your fingers above and behind your head, on your right side," Buck continues, "but squeeze down on the spoke with your thumb. Typically, when beginners first throw this loop, they will let the spoke go with the loop, and the loop will fall flat. It will wrap around the cow's legs, but it will not stand up and will not catch anything."

The loop's tip will curl in front of the cow's hind legs so that when she steps forward, her hind feet will be inside the loop.

"As you deliver the shot, turn your hand over to the dally position in the event the cow leaves quickly," Buck adds.

Take your slack as the cow moves away from you, backing your horse until the rope comes tight around the cow's legs. Push your coils forward, then dally, but only when all the slack is out of the rope. If you dally too early, with too much slack in the rope, the loop will fall from the cow's hind legs and she will escape. Once the cow is caught, keep your horse faced up to her.

Scenario 7. The overhand heel trap travels in front of the cow's right hind leg, opening so that when the cow steps forward, she is heeled.

7A.

Working Scenarios

7B.

7C.

7D.

8. The cow travels from the roper's left to the roper's right; a head catch is needed.

Potential throw: the houlihan
Build your loop and begin a houlihan swing with your thumb down, fully extending your right arm behind you.

Speed determines the angle of the houlihan swing. The faster a cow travels, the more vertical your loop should be. A fast-moving cow requires a loop that is straight up and down, while a cow standing still calls for a nearly flat swing.

After the cow passes in front of your horse, throw the loop with your arm fully extended, follow through with your right hand, and when you have made your catch, pull your slack back past your hip.

If a houlihan is thrown well, it will often form a figure eight, catching a front leg in the head loop. This can be advantageous, keeping the cow from choking down while you wait for a heeler to make his catch. And if the heeler happens to catch a single hind foot on the same side of the cow, you can easily lay the animal down.

Scenario 8A. With sidearm, overhead, and houlihan throws, a roper can send a loop farther by throwing coils with the loop. "It is like knife throwing," Buck explains. "Judge your distance, 'drop' what you think is an appropriate number of coils, and reach through the coils to hold the loop." Here, Joel has dropped coils in his poly rope in preparation for making a long-distance throw into the rodear, which is gathered in the corner of the pasture.

Scenario 8B–F. In this sequence, Buck uses a houlihan to make a long-distance head catch. When you release a long-distance houlihan, you will put more energy into your swing to create a stronger slingshot effect, and when you release the loop, you will push it with the heel of your hand. On longer shots, you will not worry about trying to create a figure eight in the loop. Instead, as the rope travels a longer distance, you will have time to take hold of the spoke so that when it settles around the cow's head, you can "set the hook," pulling slack out of the loop. "Otherwise," Buck points out, "the cow will run right through your loop."

8B.

8C.

8D.

8E.

8F.

9. **A cow travels from the roper's left to the roper's right; a heel catch is needed.**

Potential throw: the flank shot
Begin with a backhand swing, angling the loop so that it travels inward as it passes overhead and outward as it travels past your heel. Ride forward so that your horse passes behind the cow, traveling on a path about six feet from the cow's hind end.

As your horse passes the cow's tail, make your throw, sending the tip of the loop in front of the cow's left hind leg as you hold on to the spoke with your little finger. (Tip: Try splitting the spoke and loop with your little finger. This gives you more control, allowing you to hold the spoke and release the loop.)

The loop should open up in front of both hind legs, setting the trap shot.

"With the backhand flank shot," Buck says, "you can throw early, reaching over the top of the cow, and still deliver, wrapping the loop around the cow's hind legs."

When the cow steps into the trap, pull your slack, cue your horse to face up, and dally.

Safety Point: After you have made a head catch, try to keep the length of rope between your dallies and the captured cow as short as possible. The less rope that is played out, the less chance there is for the rope to tangle on objects, the heeler's horse, or other cattle.

On a related note, the heeler should never ride a horse in the area between a header and a cow that has been roped. The header's rope could easily become wrapped around the heel horse's legs, leading to an almost certain wreck. Draw an imaginary circle around the head horse. Since you do not know where the cow will travel while she is on the end of the header's rope, avoid this entire area.

On smaller cattle, the header will "log" off, keeping a calf moving straight ahead. On bigger cattle, the header will let a cow stand.

Coiling on the move
So that excess rope is not played out, and to increase control of a cow and make a heeler's job easier, the header should take every opportunity to "get short" on a cow, decreasing the amount of slack between the cow's head and the dallies around the saddle horn.

To do this, the header will remove the dallies, ride toward the cow to create slack in the rope, coil up, redally, and repeat the sequence, gradually decreasing the distance between the horse and the cow that has been headed.

This means the header will need to coil extra rope while walking toward the cow; this is called *coiling on the move*. When doing this, it is important not to walk forward faster than you can coil your rope. If you let slack build on the ground between your horse and the cow, you create a dangerous situation in which your horse might become tangled in the rope. Be sure to recoil with your right hand in dallying position.

Scenario 9. To set a backhand flank shot, a roper makes a backhand swing, and as he rides past the cow's back end, sends the tip of the loop in front of the cow's left hind leg. With the loop open in front of the cow's hind legs, when she steps forward, the heel catch will be made. Here, the cow has stepped forward with her right hind leg, which is now inside the loop, and is a moment away from bringing her left hind leg into the loop. At that point, a rider will pull the slack—tightening the rope—cue the horse to face up to the cow, and dally onto the saddle horn, or ride off and dally, depending on the rider's and the horse's abilities.

Three-Man Doctoring

A three-man doctoring team consists of a header, a heeler, and a "doctor," the cowboy who will be responsible for the necessary groundwork, which might involve a vaccination or other treatment. (The doctor is generally the buckaroo who did not make his catch.)

First, the header makes a catch and dallies. Then, he gets short by popping his dallies, maneuvering closer to the cow to draw in slack, then redallying. If necessary, the other two riders can push the cow toward the header, making it easier for the header to draw in slack.

Next, the heeler makes his catch and dallies. With some heel shots, the header might have to drag the cow forward so she will step into the heel loop. It might be necessary for the heeler to get short, just as the header did.

With the header and heeler facing the cow, both slowly back their horses, tightening their ropes. The cow stretches, becomes unbalanced, and drops to her side. Often, if experienced, the header will be facing away until the cow is down.

The doctor then dismounts, approaches the cow, puts one knee on her neck, and bends back her top front leg. The header rides forward to put slack in the rope, and the doctor removes the head rope from the cow. If only one hind leg is caught, the man on the ground always puts the heel rope around both hind feet before working with the head rope.

Then, the doctor places the head loop around the cow's top leg, reaches through the loop, bends the bottom front leg at the knee, and adds the bottom front leg to the head loop. The header pulls the head rope tight again, and the cow is immobilized for doctoring.

Once the ground man has finished his work with the cow, he steps out of the way, returning to his own horse. The header and heeler ride toward the cow simultaneously. Both pop their dallies by lifting straight up with their right hands. Each rope now has enough slack for the cow to easily struggle free, stand, and walk away, leaving the two empty loops on the ground.

An alternate strategy involves the header and heeler making their catches while the third man remains in the saddle. The heeler gets short, then rocks the cow so that her feet are off the ground. The third cowboy then rides toward the cow, drapes a loop around her two front feet, and pulls his loop closed. He then backs his horse and dallies on, slipping rope as needed as he essentially takes the header's position. The header then dismounts, removes his head rope from the cow, and doctors her. ❦

1. A team of three cowboys makes for an effective "doctoring" team. One roper heads the cow; a second roper heels her. The two then slowly back their horses, taking the slack from their ropes and gradually throwing the cow off balance so that she falls to her side. From that point, the third rider, the "doctor," dismounts to vaccinate or otherwise treat the headed and heeled animal. Here, Joel (in the foreground) holds his heel rope in place, and Roland Moore (in the background) does the same with his head rope. As his horse stands ground-tied nearby, Buck doctors the cow on the ground.

2. Safety is always the priority when working with livestock, and a "ground man" has to stay sharp to avoid an injury. Here, Kevin has just "tailed down" the cow. Next, with help from the header and heeler, he will reposition the heel rope so that it is around both hind feet, immobilizing the cow so that she can be doctored. Kevin's McCarty is still in place, tucked into his belt, and his horse stands quietly nearby.

3. In this three-man doctoring scenario, the cow was headed and heeled, then both ropers slowly backed their horses, tightening the ropes until the cow became unbalanced and dropped to the ground. Here, the "doctor," Roland, has dismounted and removed the head loop in order to place the loop around the cow's front legs. As he works on the ground, the heeler keeps tension on the heel rope and ensures the cow's hind feet stay off the ground. To prevent the cow from throwing her head and potentially injuring him, Roland keeps a knee on her neck.

4. An alternate three-man strategy involves the third roper draping his loop around the cow's front feet after she has been headed, heeled, and dropped to the ground. The third roper will then pull his loop tight, and the header will dismount to doctor the cow.

Three-Man Doctoring

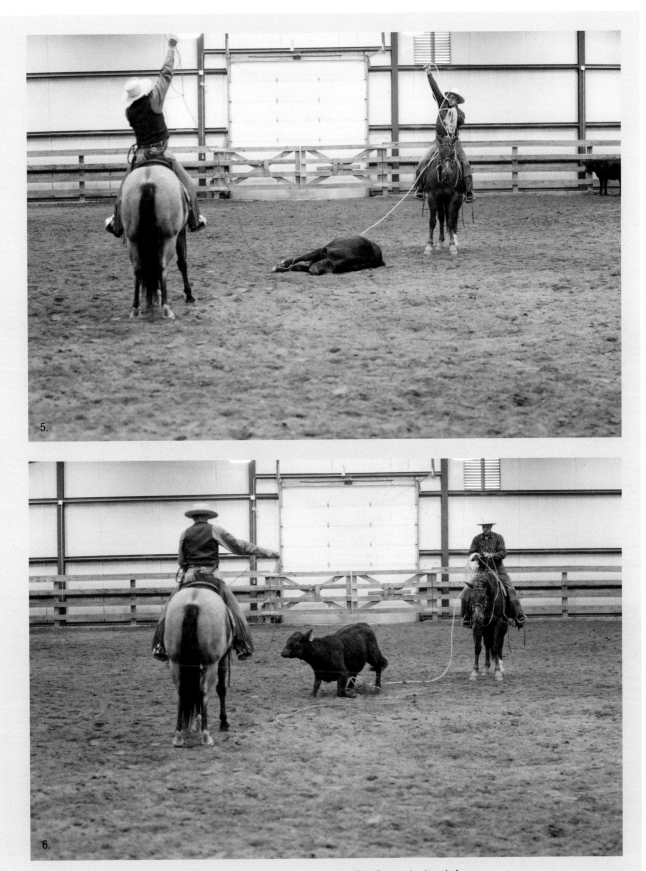

5–6. In unison, Kevin and Buck pop their dallies, putting slack in their ropes to allow the cow to struggle free.

TAILING DOWN: In instances when the cow remains standing after she has been headed and heeled, the ground man will have to *tail her down*, pulling the cow over by her tail. Here, Buck outlines how this is done:

"When you go to tail down a cow, you want to tail her down from the side," he says. "If just one hind foot is roped, work from the side the foot rope is on so you can get her off balance, on the ground, and rolled over."

As you take hold of the cow's tail, both the head rope and heel rope will draw tight, but you will still need about a foot of slack from the header, who should slip rope, feeding that slack toward the cow so that she can get her head to the ground just as you pull her tail.

"If the header does not slip rope, he is fighting you," Buck points out, "keeping the cow from getting to the ground. This is another good reason for a slick horn on your saddle: so you can run a little rope and take a cow down easier."

As the cow falls to her side, her off-side hind leg (now, in effect, the top hind leg, as the cow is on her side) will tip up in the air, at which point the ground man can take hold of that leg, using a backward hand and arm motion.

1–3. Here, just one hind foot has been roped. Buck works from that side to tail down the cow, getting her off balance and on the ground.

Three-Man Doctoring

4. Buck takes hold of the cow's hind leg, using her tail to help keep her "rocked up."

5–7. Keeping the cow's hind legs immobilized, Buck repositions the heel loop so that it is around both hind feet.

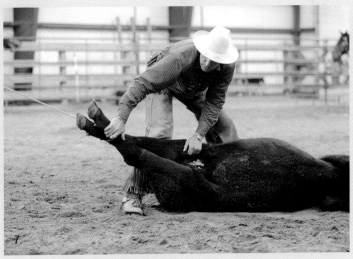

"Since a cow gets up hind feet first," Buck explains, "you have got to keep her hind feet slightly elevated, off the ground. There are two ways of holding the cow so she is 'rocked up.' I can tuck her tail under her upper hind leg and pull on the tail to keep her legs off the ground. Or I can just take hold of the hock on that upper leg and tip her up. Either way, you have to be careful not to get kicked, but a cow has very little strength when it comes to kicking forward. That is why you hold the top leg back."

Sticking with the scenario in which the cow has been roped by just one hind foot, at this point the ground man will reposition that heel rope so that it is around both hind feet, crossing the cow's hind legs in the process so that there is less chance of the cow breaking one of her legs. Then the ground man turns his attention to the cow's front legs, holding the top front leg back, with his knee placed against the cow's neck.

"Once the cow is tailed down, she is rolled over on her side and both of her hind feet are roped," Buck says, "you move straight to the cow's front end, scoop up that front leg, and pull it back."

"With some slack in the head rope, I will remove the head loop from the cow's head, keeping my foot back so that if the cow slings her head, she will not injure me. Then I will put my knee back on the cow's neck and put the rope over the top leg, which I still have hold of. I will reach through the loop to take hold of the lower leg and bring it into the loop."

From this point, the ground man lets go of the rope and the header takes over, tightening the loop around the cow's front feet. As the rope comes tight, the ground man ensures that the cow's front legs cross, making for a more secure hold for the head rope and helping reduce the risk of the animal breaking a leg.

"When I step away from the cow," Buck says, "I put my hand over her eye, and slide my forward foot (the foot nearest the header) out of the way, so that if she slings her head, she will not smash my foot."

In any doctoring situation, the ground man, header, and heeler need to use their best judgment in gauging how tough a cow will be to manage on the ground.

"Throughout the process, if a cow's really on the fight," Buck says, "it might be best to get back on your horse before you let her up."

8–9. Next, Buck moves to the cow's front end, holds back her top front leg, and removes the head loop.

Three-Man Doctoring

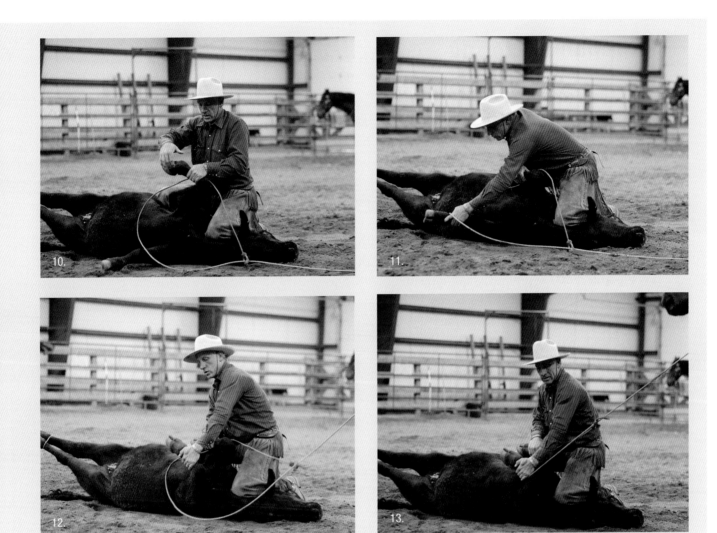

10–13. Buck repositions the head loop so it is around both of the cow's front legs.

14. Buck prepares to step away from the cow. He puts his hand over the cow's eye, and in case the cow slings her head, slides his left foot out of the way.

Two-Man Doctoring

In a typical two-man doctoring scenario, the header and heeler make their catches, the header gets short, and the two riders face up, positioning their horses toward the cow so the two cowboys face each other.

Each backs his horse slowly, so that the two ropes tighten, throwing the cow off balance so that she drops to the ground.

Then the heeler gets short, keeping the cow's hind feet off the ground as he works his way closer to the cow. Once the heeler is in position, short, and dallied again, the header works closer to the cow, continuing to get short until his horse is about six feet from the cow's head.

The header then dallies, usually taking two and a half wraps, and then steps down with his coils, leaving the reins draped over the base of the horse's neck. With the horse's McCarty tucked into his belt, the header can control his horse from the ground. This is an especially important safety point when working with a green horse.

Now the horse must be tied off to ensure the rope stays dallied while the header doctors the cow.

To do this, the header dismounts, keeping hold of the dallies with his right hand, and keeping hold of the coils with his left hand, just as if he were still in the saddle. He will leave the reins draped over the horse's neck. The header's McCarty lead rope should be tucked into his belt.

Next, the header will set the coils on the ground, ahead and to the left of the horse. This keeps the coils out of the way as the header works and helps ensure that his horse will not step in the coils and become tangled in the rope.

With his left hand, the header takes the top coil from the stack, creating some slack between his right hand, which still holds the dallies in place, and the stack of remaining coils.

The header then folds this section of rope into a simple U shape, without letting the rope cross over itself. Then, he brings the rope under then back over the taut head rope, just under the horse's neck.

1. In this variation on a two-man scenario, Kevin has headed the cow and Buck has heeled her.

2. Buck uses two half-hitches to secure his heel rope.

Two-Man Doctoring

3. Next, Buck positions his knee on the cow's neck and holds back her front leg.

4. Working carefully, he removes the head rope, keeping his knee in position on the cow.

5. Buck then brings the cow's top leg into the head loop . . .

6. adds the lower leg . . .

7. . . . and crosses the legs as Kevin pulls the slack and tightens the head rope.

8. The cow is now restrained for doctoring.

Finally, the header will bring the folded rope back to the saddle horn, where he will tie it off with two half-hitches. This should keep the dallies secure and in place as the header begins doctoring the cow on the ground.

Once the header has tied off, he puts one knee on the cow's neck and bends back her top front leg. Next, he cues his horse forward with the head rope, removes the head rope from the cow, puts the top front leg inside the loop, and then adds the bottom front leg, making sure the legs are crossed.

The header then tightens the loop and cues his horse to back gently, ideally with nothing more than light cues from the lead rope. With the cow tied and stretched, the header can doctor her.

When it is time to release the cow, the header unties the half-hitches on the saddle horn, removes the wrap he made around the taut head rope, recoils the rope while still on the ground, then remounts with the reins and coils in his left hand.

Then the header and heeler ride forward, putting slack into their ropes, pop their dallies in unison, and allow the cow to struggle free, stand, and walk away. ✥

9–12. Here, Buck has heeled the cow by one leg and he is "tying off" to keep the rope secured to the saddle horn so that he can work on the ground. To tie off, he has brought his coils with him, leaving the dallies in place around the saddle horn. Keeping a good grip on the rope with his right hand, Buck ensures the dallies stay in place as he folds the rope's slack under, then over, the outstretched section of rope running between the horn and the cow. Buck then brings that folded section of slack back to the saddle horn where he ties it off with two half-hitches. Next, he will put his remaining coils on the saddle horn, followed by his reins and romal. If the heeler has caught one foot, he steps off to tail down the critter. That way, if things go awry, the header is still in charge.

Two-Man Doctoring

13–17: Next, Buck tails down the cow, cues his horse forward to put slack in the heel rope, and repositions the heel rope so that it is around both hind feet. Finally, he cues his horse back to remove slack from the rope.

Two-Man Doctoring

18. The doctoring done, his half-hitches untied, and his rope organized, Buck remounts.

19. Kevin and Buck pop their dallies, which will put slack in the head and heel ropes, allowing the cow to struggle free and to her feet.

GROUND CONTROL: An on-the-ground handler must be able to move either end of the horse—the front quarters or the hindquarters—with a subtle suggestion.

To build this skill, Buck uses a set of ground exercises. When he steps toward the horse's hip, the horse should step away with its hindquarters, crossing its near-side rear leg in front of its off-side rear leg. Simultaneously, the horse should bring its front quarters toward Buck.

"The horse has to be able to untrack his hindquarters—to take his hindquarters to the outside, and his front quarters to the inside," Buck explains. "This will help me be able to move either end of the horse while I am on the ground."

"The horse needs to follow a feel. One of the first things I like to do with my horse on the ground—to get him ready to work a rope—is to lead the horse by me, doing all the groundwork I would teach any young horse."

For this exercise—leading a horse by—Buck starts at a standstill. First, he asks the horse to walk past him, using only pressure on the halter.

"That is no different than what I would do to get a colt ready to be ridden the first time," he says. "I need to be able to lead him by me without having to swing the tail of the lead rope, or do much else, because when I am on the ground working a cow, I will not be able to swing the tail of the rope to ask my horse to move."

Once Buck is comfortable with the way a horse leads by him, he "sends the horse through" in each direction, asking the horse to lead by him, then turn to face up. From here, the horse must learn to follow lead-rope suggestions to back so that it can be relied upon to keep a rope tight and to keep the loop closed around the cow's feet while the handler doctors the cow on the ground. Slack in the rope can create a hazard, allowing the horse or ground man to get tangled, and it can lead to the cow kicking free of the loop.

"When a horse will lead by me in both directions with a soft feel, face up to me, and then back and tighten the rope," Buck says, "I am ready to doctor cattle in the pasture, working on the ground with the horse on the end of a McCarty."

To cue his horse to step forward while he is on the ground, Buck pulls on the rope dallied to the horn, applying light pressure. At first, he reinforces his suggestions by also asking with the McCarty's lead rope. A finished horse learns to distinguish a pull on the rope from a cow and a longer, more gradual pull on the rope from a handler on the ground. ✤

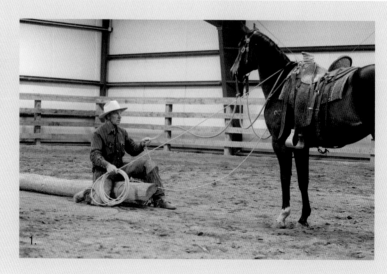

1–4. While he is on the ground, Buck can use signals from his rope and his McCarty lead to ask his horse to step back or move forward. This arena session offers good practice for doctoring sessions in the pasture.

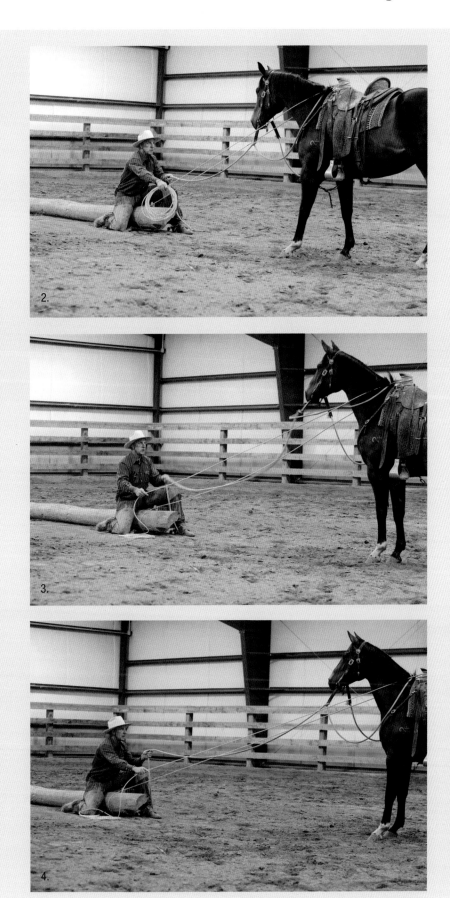

5–7. Buck prepares his horse for a day of work with some basic groundwork, akin to the steps he would put a colt through before a first ride. He asks the horse to walk past him, then untrack his hind end, then bring across his front end, all in response to lead-rope signals.

5.

6.

7.

Strategies for Beginning Ropers

A lot of people are interested in roping, but fewer are interested in the horsemanship. It's because horsemanship is hard, it's difficult to come by, and you have to work at it. You have to pay a price. But if you're a good horseman, you're apt to be a good roper.
—BUCK BRANNAMAN

The biggest mistake beginning ropers make is that they do not practice hard enough on the roping dummy or on their horsemanship. Their horsemanship might be really primitive, even crude. Then they will try to rope and they will not be safe. They will have bad experiences roping if their horsemanship is poor.

Roping is a progressive thing. Within just a few days, if you have a gentle horse and he is maneuverable and respectful and not afraid—and if you are really into it—you can be throwing breakaway shots, practicing dallying and knot roping. But err on the side of safety; go slowly. Once you have thrown a thousand breakaway shots and dragged a log a hundred miles, maybe you are ready to move forward.

A lot of times, you can find opportunities to rope with someone who is a good hand and who has gentle cattle tailored for a green roper, cattle that are not too wild, that will not punish you too much, and that will not rope-burn your hand if you are a little late getting to the horn and dallying.

If a person is in the market for a ranch-roping horse, you first look at your basic standards of conformation for a performance horse. Personally, I like a horse that is about sixteen hands tall and about 1,350 pounds, stout enough that I can rope big cattle and the horse can handle it. If someone asked me for advice on buying a horse, I would recommend they look for one that has had a couple thousand calves roped on him, and maybe four hundred or five hundred yearlings caught outside in the pasture on him. He will be a well-broke horse, and handle well. That kind of horse might cost a good deal of money, but he is cheap compared to what a trip to the hospital might cost if you are roping on something that is not prepared.

You need to operate that horse as if he were your legs, as if he were an extension of you. There are a hell of a lot of things that can go wrong, that can melt down in a hurry, when you start roping. One of the most common mistakes is taking your eyes off the cow. When you have a cow roped and you look down at your horn to dally, that is the beginning of the end.

You need to be maneuverable and accurate with your horse. Gradually work your way forward as a roper. You will get to where you can work a little faster and still be smooth. You will start out slow and smooth. Much later, you might get fast and smooth.

Know your horse and learn to read how he is dealing with a situation. A scared horse is the most dangerous variable in roping. When you dally on, for instance, your horse might get afraid and whirl away, wrapping both of you in the rope while there are dallies on the horn and a cow on the end of the rope.

When you are green at anything—skiing, basketball, shooting pool, roping—you should surround yourself with the most talented people you can find. More than likely, you will adapt to your environment and rise to their level. But if you are a green roper and you surround yourself with people that are almost as incompetent as you, you are looking for trouble. If you surround yourself with good horsemen and good ropers, they will want you to succeed.

A good roper helps his partners in a proactive way rather than in a reactive way. He helps them stay out of trouble and can give them advice in a timely manner, before they get into wrecks, before they learn the wrong way to do something. No one needs a critic who will tell you after the fact, 'I saw that coming.' That does not do any good."

There are a lot of things a beginning roper can work on at home, even if he does not have cattle. At one of my roping clinics, I met a guy from Germany. He had never roped a live animal in his life; it is illegal in Germany to rope a cow. He was quite a renowned dressage instructor, and

Strategies for Beginning Ropers

The smartest strategy for a beginning roper: Surround yourself with the most talented cowboys you can find. Skilled working partners help keep working situations under control and can prevent wrecks before they occur.

using his dressage horses, he had been roping on a dummy. He had done nothing but practice in his arena during his spare time, but when he came to that first roping clinic, he could throw all the shots I could throw, he could take his slack, and he could dally. He looked like he had some experience on a ranch, but it was really the first time he had roped cattle in his life. Depending on your level of devotion, there is a lot you can do at home, even in your backyard. And that is really cool.

At some point, a beginning roper needs to go watch good ropers on ranches or at major ranch-roping events. When you watch ropers with advanced skills, you notice things that will make you better.

If you have a skilled roper roping with you, that will keep things slow and under control. He will tell you what is going to happen before it happens. He will set up shots for you, tell you what kind of shot to throw, and talk you through throwing your shot, handling your slack, handling your horse, going to the horn, and dallying. And, if you get in trouble, he will even talk you through getting out of trouble.

Chapter Four: Tools for Training

The most important thing is that this tradition of handling reatas and horses is preserved and continued and passed on to younger people.
—BUCK BRANNAMAN

Headgear	111
Hobbling	121

Headgear

Author and Western artist Randy Steffen, among others, referred to Spanish California as the "land of mañana," a culture in which "there was always tomorrow." Horsemen believed in taking their time in training young horses and in advancing in small, modest steps.

That mind-set is not so common in the contemporary stock-horse culture, in part because of the prescribed timetables dictated by competitive events—horses must have achieved a certain level of proficiency by the time they are two years old, reached another predetermined level

A bridle horse, one that has progressed through the training phases of the snaffle bit, hackamore, and two-rein, carries a spade or half-breed bit. In keeping with the Spanish California tradition, a bridle horse wears a *bosalita,* or pencil bosal, as a matter of "proper dress" and tradition. Connected to the bridle is a set of rawhide romal reins. The bit, rein chains, and the reins' rawhide buttons all contribute to the balance of the horse's headgear. There are many other serviceable mouthpieces, but the spade and the half-breed are the classic types.

by age three, and so on—and in part because of a general unwillingness to invest time, and therefore money, in a process that can take several years.

In the buckaroo culture, though, the old California traditions have remained in place. Many horsemen in the West Coast and Great Basin ranch cultures have adhered to the vaquero mind-set and methodology when it comes to training bridle horses. Their approach has generally resulted in horses more refined in their handling than those that have been rushed into the bridle through abbreviated training regimens.

Today, thanks to the work of traveling clinicians, such as Buck, and the many forms of equestrian media that now exist—books, magazines, CDs, DVDs, horse-oriented television networks—more horse owners are becoming aware of California bridle-horse traditions, recognizing their benefits, and putting them to use in their training programs in the interest of good horsemanship for the sake of good horsemanship.

Here, we offer a brief overview of the bridle-horse progression. ❧

From left: snaffle bit rig with McCarty; a hackamore; a two-rein, including a half-breed bit with an under-bridle hackamore; and a spade bit with a bosalita and neck rope, for leading the horse. In the California tradition, a horse is started in the snaffle bit, which operates on direct-rein cues, then advances to the hackamore, in which the horse develops suppleness and more refined responses to the rider's reins and legs. The horse then progresses to the two-rein, carrying both a bit (a spade or half-breed) and hackamore. Gradually, the rider transitions from asking with the hackamore to asking with the bridle bit, until the hackamore is no longer necessary. Throughout all phases of the horse's education, the rider's legs should always be in harmony with what the rider's hands ask.

Headgear

While Kevin is on the ground, a McCarty tucked in his belt gives him a link to his horse and provides him with a means of controlling his horse as he works. Buck prefers a McCarty with no knot tied in the end, so that the lead rope can easily pull from a cowboy's belt. This is a safety issue, ensuring the cowboy will not be dragged by a panicked horse, and a matter of convenience for the cowboy, allowing him to more easily remove the lead rope.

THE MCCARTY: Anglicized from *mecate*, the Spanish word for "rope," a McCarty is essentially a section of rope, often braided horsehair, that attaches to either a snaffle bit or hackamore, forming a set of closed reins and a lead rope. Should a cowboy have to dismount, the McCarty lead rope allows him to keep control of his horse while on the ground.

"If I am working on the ground with a roped cow, and the rope on the cow gets too loose," Buck explains, "I can use the McCarty lead to set my horse back and tighten the rope. If a green horse got scared, he could drag a calf to death. With a McCarty, I can keep contact with the horse. It is my connection to him."

Most working cowboys who use McCartys tuck them in their belts so they are close at hand and need not be untied from the saddle.

"I prefer a McCarty with no knot tied on the end of it so if a horse fell with me, I would not get dragged," Buck says. "And when I step off to doctor a cow, the rope will pull out of my belt freely."

THE SNAFFLE BIT: Traditionally, stock horses are started in the snaffle bit, contemporarily defined as a "broken" two-piece bit that is generally connected on each end to rings that serve as cheekpieces. In the buckaroo tradition,

Roland uses a snaffle bit on this roan colt. Attached to the rings of the snaffle are leather "slobber straps" and a McCarty.

McCarty reins are preferred when the snaffle bit is used, but many riders opt for roping reins or for split reins that have been crossed.

In the Western tradition, snaffles are generally used with leather slobber straps that link the McCarty and the snaffle bit itself. The slobber leather connects to the ring of a snaffle better than the McCarty, which again is essentially a rope.

The snaffle is a common bit and one that, generally speaking, all horses and riders can be comfortable using. Its mechanics are simple: the snaffle lacks shanks, so there is no leverage to speak of, and the cue is direct; the rider applies pressure with the hand, and that pressure acts on the corner of the horse's mouth.

When using a snaffle bit, a rider rides with two hands. To ask a horse to turn to the left, the rider, with his or her right leg positioned forward, pulls lightly on the left rein; right-leg pressure reinforces the rein signal to turn to the left. To ask the horse to turn to the right, the rider, with his or her left leg positioned forward, pulls lightly on the right rein; left-leg pressure reinforces the rein signal to turn to the right. If the rider positions his or her legs farther back, these signals will, in effect, ask the horse to move its hindquarters rather than its front end.

To help prep a horse to one day work from neck-rein cues, the rider can supplement a direct-rein cue for a turn by placing the opposite-side rein against the horse's neck. For instance, while asking for a left turn by pulling on the left rein, the rider can also lay the right rein against the horse's neck, allowing it to begin associating that sensation with a turn to the left.

Once through the awkward initial rides of the colt-starting phase, a horse will begin to find comfort in yielding its head, lowering its head and neck, and flexing at the poll.

Many horses start in the snaffle and stay there. Advancing beyond the snaffle in a way that benefits the horse and rider takes a concerted effort and time spent focusing on that specific goal. That can be a luxury on a working cattle outfit, where a cowboy might have several horses in his string, all somewhat green, and not enough hours in the day to do his work and further develop a horse beyond the snaffle-bit stage, though many cowboys have a great interest in fine horsemanship and have the horse-handling talent to make fine bridle horses. ❧

THE HACKAMORE: Once a horse has succeeded in the snaffle bit, it graduates to the hackamore. The hackamore horse performs the same work as it did in the snaffle bit but becomes softer and more refined in its handling as a result of the quality time spent.

In the bridle-horse tradition, the hackamore, being without a bit, can have a strategic role, offering an alternate means of communication and control at a stage in a horse's life when it is shedding teeth and therefore has a particularly sensitive mouth. Plenty of horsemen contend that colts should be started in the hackamore rather than in the snaffle, and that there is no reason for a horse to carry a bit before its mouth matures.

A hackamore consists of the headstall, McCarty, and rawhide bosal, or noseband. Many horsemen commonly refer to the noseband itself as the hackamore. Still others make a distinction, calling larger-diameter nosebands hackamores and smaller-diameter nosebands bosals.

Constructed of braided rawhide, a bosal also has a rawhide core, which offers the benefit of easily conforming to the shape of an individual horse's jaw or muzzle.

The hackamore's mechanics are unique. Altogether lacking a mouthpiece, it applies pressure to the horse's jaw. Unlike the snaffle, which applies direct pressure to the horse's mouth on the same side as the rein in use, the hackamore applies pressure on the side opposite of the rein in use. A pull on the left rein brings the right

A hackamore includes the McCarty, headstall, and rawhide bosal, which is a noseband of braided rawhide around a rawhide or nylon core. Unlike the snaffle, which operates on direct pressure, the hackamore works by applying pressure to the horse's jaw on the side opposite the rein being pulled. In the hackamore, a horse becomes suppler and achieves greater refinement in its responses to rein signals. As a rider achieves more refinement with a hackamore horse, he or she transitions from using heavier, stiffer bosals on the horse to lighter, softer bosals.

Time spent in the hackamore teaches a horse to tip its nose in and break at the poll rather than at the withers, provided the horse is released while in the proper position. However, the notion held by some horsemen that the hackamore makes a horse better than it could have been in the snaffle bit is not true. The hackamore complements the work that has previously been done in the snaffle.

In the bridle-horseman's philosophy, a horse stays in the hackamore for as long as it benefits the horse and for as long as the hackamore offers continued improvement. Generally, as a rider achieves more control with a hackamore horse, he or she moves from heavier, stiffer bosals to lighter, softer bosals.

A bosal should be positioned about halfway between the horse's eye and nostril and just above the wide spot on a horse's nose so that the noseband lies against bone rather than cartilage.

A hackamore's headstall is often positioned so that it runs just behind the horse's eye—so close, in fact, that observers might reasonably assume this position is unsafe or uncomfortable for the horse. Nevertheless, such positioning is normal and should neither cause discomfort for the horse nor create the risk of injury to its eye. The anatomy of a horse's skull is such that the headstall should not slip forward over the eye.

A more sophisticated piece of headgear than the snaffle, the hackamore requires the rider to work with a good degree of "feel," an innate instinct of how a horse will respond to pressure from the rein, the hand, and leg. When a hackamore horse responds to a signal, the rider must immediately recognize the response is occurring and release pressure, helping the horse learn to respond correctly to cues through positive reinforcement.

Savvy horsemen contend that the hackamore is vastly underrated, and that it is in the hackamore that a horse learns the fundamentals of being a bridle horse. ✤

side of the bosal into contact with the horse's jaw; a pull on the right rein brings the left side of the bosal into contact with the horse's jaw.

THE TWO-REIN: When a hackamore horse is ready to advance to the next stage of bridle-horse training—carrying a single-pieced, shanked bit—vaquero tradition dictates a two-rein stage in which the horse remains in a small-diameter pencil bosal and begins carrying a bit, typically a spade bit or a half-breed, which is a variation on a spade.

The spade's most recognizable feature is the spade-shaped port that touches the horse's palate and tongue, sending signals to both. The bit includes a rolling "cricket" made of copper, an element horses like the taste of, and that triggers the horse's saliva glands, moistening its mouth and, in effect, lubricating it, making the bit more comfortable to carry. Brace wires provide stability for the spoon and prevent lateral twisting.

The spade's complex and imposing architecture has made it one of the most misunderstood bits; the uninitiated dismiss it as harsh, even cruel, basing their conclusion on the bit's length and the height of the spoon.

As Oregon bit-and-spur maker Ernie Marsh once noted, though, "If your tongue was a foot and a half long, a four-inch spoon would not scare you so much."

The spade actually fits easily in an adult horse's mouth and is meant to be used by the most advanced riders, those capable of keeping their hands light and of signaling their horses with subtle, almost imperceptible cues given with slack in the reins. Often, a bridle horse responds to cues as light as the swinging of a rein chain or a slight change in the pressure of a rein against its neck.

In the days of Spanish California, half-breed bits were often used by riders whose horsemanship was not advanced enough to handle the spade correctly or for horses that did not respond well in the spade or would not accept it. It no longer has a stigma of inferior horsemanship and is a well-respected communication tool.

At this stage, as the two-rein label would imply, the rider uses two sets of reins: a McCarty, connected to the bosal, and rawhide romal reins, connected to the bit.

Romal reins are typically around forty-two inches in length, and are linked to the bit by rein chains, which help keep the bit itself balanced and in place inside the horse's mouth and serve as buffers for rein cues. When a rider begins to exert pressure on the reins, the signals must travel through the chains, limiting hard jerks on the horse's mouth. Chains also help protect the rawhide reins should the horse need to take a drink while bridled.

For romal reins to work effectively, they should weigh as much as the rein chains. To add weight to the reins, rawhide braiders line them with rawhide "buttons," which help balance the reins' weight against that of the chains and bit.

The romal itself, which is essentially a rawhide quirt that is attached to the reins and held in the rider's right hand (when not roping), is heavier than the reins, and it too is lined with rawhide buttons. On the end of the romal is a leather popper, usually around twelve inches in length.

Early in the two-rein stage, the rider uses the hackamore to control the horse, sending signals that the horse will have already become familiar with. The bit—again, either a spade or half-breed—is simply there so the horse can learn to carry its weight and become accustomed to its presence.

Gradually, the rider begins asking with both the hackamore and rawhide reins, then begins weaning the horse off the McCarty signals altogether to the point that the bosal is no longer necessary. As he makes this transition from emphasizing the hackamore to emphasizing the bridle-bit reins, the rider will hold the McCarty reins at longer and longer lengths, over time decreasing their influence over the

In the two-rein phase of bridle-horse training, a horse carries a light bosal and begins packing a bit—either a spade or half-breed, which is a modified spade. The rider carries a set of reins for each piece of headgear—romal reins for the bit, a McCarty for the bosal. At first, the rider will signal with the McCarty and bosal, allowing the horse to become accustomed to carrying the bit. Gradually, the rider will begin incorporating signals from the reins and romal and, over time, wean the horse from the hackamore signals altogether so that it works from rein signals alone.

Headgear

horse as the bridle's influence grows.

When it comes to roping, a rider holds the romal reins, the McCarty reins, and the coils of the rope in the left hand and drapes the romal over the horse's left side, freeing the right hand to handle the loop and dally. ✣

The two-rein presents a unique set of challenges for a roper, particularly when it comes to handling the McCarty, reins, coils, and dallies. Here, Joel has both sets of reins and his coils in his left hand as he holds the dallies with his right.

A rider never leads a bridle horse by the reins; doing so would create pressure on the bit. Instead, a lead rope provides for control of the horse on the ground. Here, the lead rope is secured around the horse's neck. Buck holds the lead rope draped over his arm while he is on the ground. Note that the bosalita hangs from a leather strap tied into the horse's forelock.

THE BRIDLE: In the final headgear phase of bridle-horse progression, the horse works exclusively from rein signals sent to the bit—again, either a spade bit or a half-breed variation—in conjunction with the horseman's legs.

In keeping with tradition, the bridle horse still carries a pencil bosal, although at this stage of training it plays no role in signaling. The bosalita is a show of respect to the bridle horse.

When a rider has dismounted to work on the ground, he or she should never use the reins to lead a bridle horse, as this creates pressure on the bit. Instead, the rider should use a light lead rope to keep in contact with the horse.

Hobbling

"If a person is going to be any kind of a hand, he needs his horse to be hobble-broke," Buck says. "If you are at a branding and take your turn working on the ground, when the boss tells you that you can rope again, it is nice if you can find your horse."

Horsemen include hobbles as part of a horse's training often as a way of teaching the horse patience; to stand still and quietly while the cowboy has to turn his attention elsewhere on the ground, perhaps to put in his time as a ground man while others continue roping; or to repair a patch of downed fence. Hobble-training also helps a horse learn to accept restraint; in the event a horse gets a foot caught in a wire fence or other "trap," it is less likely to panic, given that its training has included the acceptance of foot restraint.

As with many training concepts, the best place to introduce a horse to hobbles is an enclosed, but spacious, area—a large round pen, for instance—with good, strong fencing. The cowboy and horse both will need room to maneuver without running the risk of getting caught against a fence.

The horse must be prepared for this by extensive work with ropes around its legs—for example, leading it by a front foot and stopping it by a hind foot while the horse is worked inside a round corral—all of which is part of basic colt starting.

A handler should begin the initial hobble-training session by rubbing the horse, getting it calm and comfortable.

With the horse standing square, particularly with its front legs, the handler can put the hobbles on the horse. Initially the hobbles go around the pasterns, then eventually around the cannon bones.

The hobbles should be adjusted so there is an inch or two of slack around each leg so that the hobbles do not cause a soft-tissue injury to the horse's legs; however, the hobbles should be tight enough to prevent them from falling down around the horse's hooves and working completely free.

"I always use a single hobble on the back leg, with a rope between the front to prevent a horse from running off," Buck says. "I do this for a few sessions before I just hobble in front."

Hobbles initially confuse, even frighten, the horse. A handler can reassure the horse by continuing to gently rub it after the hobbles have been put in place. Then the handler should keep a loose hold of the lead rope and stand by at a safe distance to give the horse a sense of security and to keep control should it try to run off with the hobbles on its legs. Keep the lead rope out of the horse's way so the horse does not become entangled in it as it first tests the three-way hobbles.

Typically, a horse will try to walk out of, or through, the hobbles. The horse might struggle against the hobbles, even rear, but should eventually accept them and stand quietly. All the while, the handler should keep a quiet hold of the lead rope and exert as little pressure as possible. Ideally, no signals should come from the handler at all, as if the handler were not present. When the horse has stood quietly for several minutes, the hobbles can be removed.

This first session will take as long as it takes: a half hour, perhaps an hour. Subsequent sessions will go faster, since the horse will already be familiar with hobbles and with the process of being hobbled.

Once the horse can be trusted to stand quietly while hobbled, the handler can put the lead rope on the ground, or drape it over the horse's back, and take a few steps away, leaving the horse standing on his own. The handler can

Hobbling

quietly increase the distance, taking additional steps. As the horse develops this skill, the handler can eventually leave altogether, confident the horse will stand quietly and be safe. At first, though, keep careful watch over the horse from a distance to be sure it is handling the circumstances well and will not hurt itself.

"I do not ground-tie," Buck explains, "but I will put the lead rope down on the ground or around the saddle horn and let the horse stand alone, after I step away."

Once a horse is hobble-broke, a cowboy can dismount, lead his horse away from the action at a branding or rodear, hobble the horse, and do his work on the ground, confident his horse is safe, comfortable, and content, and that the horse will be there when it is time to remount. An unexpected fencing chore, or another circumstance that calls for the cowboy to be on the ground, is made much easier with a hobble-trained horse.

Hobbles help ensure your horse is still there after you have finished work on the ground. When a handler first introduces hobbles, a horse can find the experience confusing and even frightening. It is important that in early hobble-training sessions a handler remain with the horse to keep it calm as it first tests the restraints. Usually after only a few training sessions, a horse begins to accept the hobbles and can be trusted to remain calm and relaxed while wearing them. Hobble-training also helps keep a horse from panicking should it get a foot caught in a wire fence.

Credits

Roping is a lifestyle—handling your horse as good as you can, handling your cattle as good as you can, being as skilled as you can be.

—BUCK BRANNAMAN

Joel Eliot is a horseman, farrier, working cowboy, and Western musician from Sonoita, Arizona.

Kevin Hall is a rancher, horseman, and farrier from Kiowa, Colorado.

Roland Moore is a rancher and horseman from Norris, Montana.

A. J. Mangum is a lifelong horseman and the editor of *Western Horseman* magazine. He and his wife live in the ranching country east of Colorado Springs, Colorado.

Buck Brannaman is a horseman, rancher, and clinician. When not on the road conducting clinics on colt starting, horsemanship, and ranch roping, he lives with his wife and daughters on their ranch outside Sheridan, Wyoming.

Training, Riding, and Caring for Your Horse

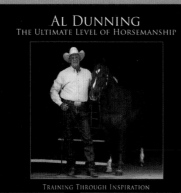
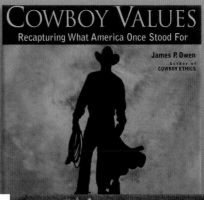

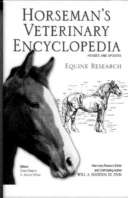
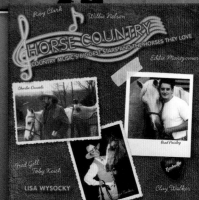

- ★ Natural horsemanship
- ★ Western and English training
- ★ Horse care and health
- ★ Dressage
- ★ Literature for horse enthusiasts

For a complete listing of all our titles, please visit our Web site at www.LyonsPress.com.

Available wherever books are sold.
Orders can also be placed on the Web at www.LyonsPress.com, by phone from 8:00 a.m. to 5:00 p.m. at 1-800-243-0495, or by fax at 1-800-820-2329.